GLOBAL ENVIRONMENTAL CHANGE

Introduced Species

GLOBAL ENVIRONMENTAL CHANGE

Introduced Species

Scientists classify plants and animals with which humans share the environment among three categories: native, introduced, and invasive species. Native species are established within an ecosystem, having evolved there over thousands of years. Introduced species have been brought to an ecosystem, intentionally or otherwise, through human activity. Invasive species are "invasive" because they disrupt ecological processes and harm human systems.

While only a minority of introduced species cause great disruption and harm, they may be found in almost all regions of the United States. Many introduced species are beneficial to humans or, at the very least, innocuous. Students may be surprised at how many "everyday" species are introduced. Almost all North American crops, livestock, and domestic and game animals, many sport and aquacultural fish, numerous horticultural products, and most biological control organisms were introduced from other continents. Introduced species are essential to North American industry, providing economic, recreational, and social benefits.

But even everyday introduced species—species which are at present considered beneficial or innocuous—have ecological impacts, and the full range of these impacts will take years to understand. While everyday introduced species may be beneficial in the short term, the long-term impacts of a small fraction may turn out to have caused some harm in the future. The human activity of introducing species from one continent to another, from one ecosystem to another within a continent, and from one habitat to another within an ecosystem, is changing environmental systems and processes on a global scale.

WHAT IS AN INTRODUCED SPECIES?

Scientists define an introduced species as any nonindigenous species found beyond its natural range. Introduced species have been inserted into a new environmental situation through human activity. While immediate negative impacts are rare, overall impacts on ecological processes and human systems can take years to become fully apparent. Scientists' ability to accurately predict such impacts is limited.

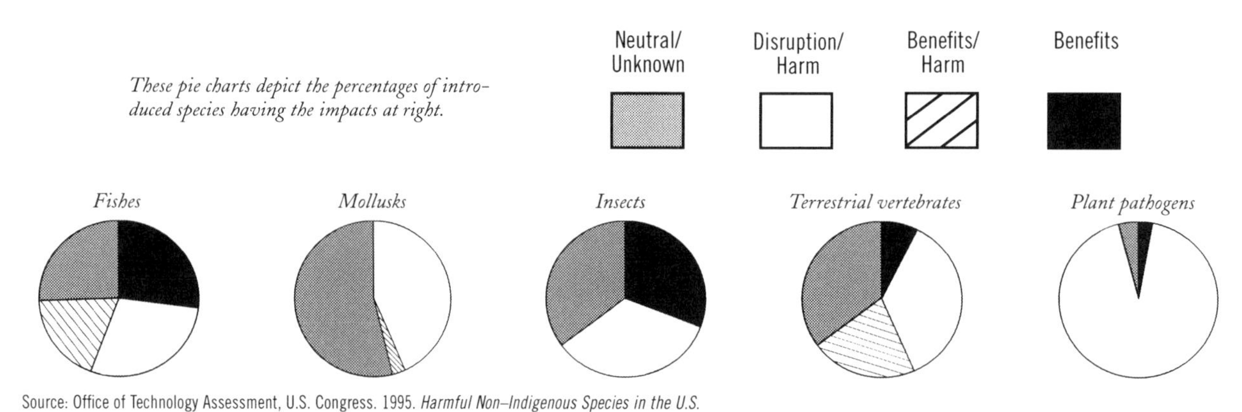

These pie charts depict the percentages of introduced species having the impacts at right.

Source: Office of Technology Assessment, U.S. Congress. 1995. *Harmful Non-Indigenous Species in the U.S.*

GLOBAL ENVIRONMENTAL CHANGE

Introduced Species

published by

National Science Teachers Association

with support from

Environmental Protection Agency

Office of Research and Development

NSTA's *Global Environmental Change* series is science based, and links the ecology and biology of global environmental changes with insights and information from other disciplines. The series teaches students how to gather a wide range of relevant data and information derived from pertinent areas of study, and encourages them to pose and then answer their own questions in order to make their own decisions about solving global environmental problems.

Introduced Species is the fourth installment in the *Global Environmental Change* series, which the National Science Teachers Association created in conjunction with the U.S. Environmental Protection Agency's Office of Research and Development. Other installments also use case studies to focus on such global environmental topics as biodiversity, deforestation, and carrying capacity.

The National Science Teachers Association, founded in 1944 and headquartered in Arlington, Virginia, is the largest organization in the world committed to the improvement of science education at all levels—preschool through college.

ACKNOWLEDGMENTS
Introduced Species' contributors included Irwin Slesnick, Linda Wygoda, Susan Plati, Richard Johnston, and Dan Simberloff. The series was developed by: Ron Slotkin, EPA Office of Research and Development; Shirley Watt Ireton, NSTA Publications, Director; Gregg Sekscienski, NSTA Publications, former Associate Editor; and Margaret Edwards, writer. Ongoing support for the *Global Environmental Change* series is provided by the U.S. Environmental Protection Agency's Office of Research and Development.

Introduced Species was published by NSTA—Gerry Wheeler, Executive Director; and Phyllis Marcuccio, Associate Executive Director for Publications—and produced by NSTA Special Publications—Shirley Watt Ireton, Director; Chris Findlay, Associate Editor; Anna Gillis, Associate Editor; Michelle Treistman, Assistant Editor; and Christina Frasch, Assistant Editor. Chris Findlay was Project Editor. The series was designed by Daryl Wakeley of AURAS Design. The cover was designed by Dan Banks of AURAS Design. David Miller was the principal artist.

The cover photograph was taken by Tony Stone Images. The back cover red fire ant inset is by Sanford Porter and the water hyacinth inset is by Tony Stone Images. The back cover map is from Digital Wisdom, Inc.'s Mountain High Maps and Globeshots CD-ROM. The *Washington Post* article, "Serpentless Hawaii Fears Snake Invasion," by William Claiborne is reprinted with permission.

PB138X04
ISBN 13: 978-0-87355-167-0
ISBN 10: 0-87355-167-2

USING THIS BOOK

The activities in this book are designed to equip students with scientific tools and skills for understanding what introduced species are, how they impact natural processes and human systems, and what may be done about them. The activities link the biology and ecology of introduced species with insights and information derived from other disciplines and areas of study.

Organizationally, the book's activities follow the "journey" of an introduced species once it reaches a new environment. All introduced species experience some variation on the stages covered by each of the activities: they disperse, they interact with other organisms, they adapt to new environmental situations, they alter ecological processes, and they compete with humans for control of environments. Each activity focuses on a different introduced species at a different stage in this journey. Use this scheme to demonstrate to students that *all* introduced species—from bacteria to fungi to plants to animals—are governed by the same organizing scientific principles.

The imported red fire ant (*Solenopsis invicta*) is used as a case study for this journey. Each activity contains a sidebar demonstrating how the red fire ant moves through that particular stage of its own journey. Use this ongoing red fire ant storyline to keep students focused, and to help them connect their general understanding of the science behind introduced species with more specific knowledge about a particular species.

A key feature of this book is the introduced species journal students maintain (some suggestions about how to organize student journal-keeping appear on page 10). Keeping journals enables students to organize their information as they gather it, to combine information within peer groups, and to integrate all the information gathered by the entire class toward successful participation in the culminating activity, Balancing Human Systems with Natural Processes.

Even a cursory database keyword search reveals a sea of detailed information about many different introduced species, and maintaining their own journals will help students keep on track. Journal-keeping will also enable students to make direct links between aspects of this global environmental topic and their own experiences. Encourage students to use the fire ant storyline, as well as the book's individual activities, to guide their research into introduced species with which they are already familiar or have direct experience.

These activities don't hand students answers, but instead provide a framework in which students may progressively build layer upon layer of knowledge. Starting with Activity 1, students will begin a process of gathering a wide range of information, and of acquiring and combining science-based with interdisciplinary tools and skills. This process will enable them to pose and answer their own questions, make their own decisions, and solve problems related to introduced species.

The red fire ant icons below appear next to the box containing the ongoing fire ant storyline in each activity.

Red Fire Ant Queen

Red Fire Ant Winged Male

Red Fire Ant Minor Worker

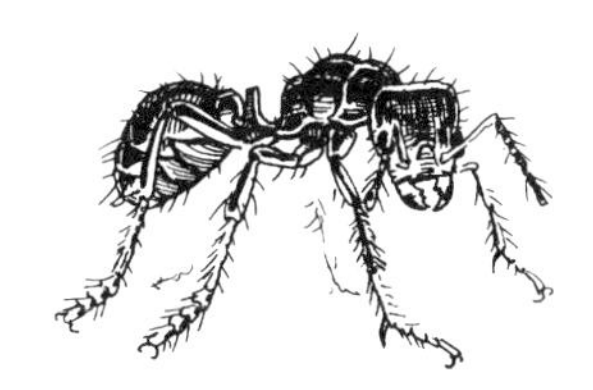

Red Fire Ant Major Worker

THE CASE STUDY

RED FIRE ANTS

Red fire ants (*Solenopsis invicta*) are native to the Pantanal region of southern Brazil, and were introduced to Mobile, Alabama, sometime around 1940. Cargo ships in those days used soil to ballast their holds for return journeys, and the soil would simply be dumped in whatever port of call was next on the ship's itinerary. In this way, many species have been inadvertently introduced to other parts of the world from their native regions. In fact, three other fire ant varieties preceeded *S. invicta* to North America. Two of them (*S. geminata* and *S. xyloni*) are considered native to North America, but they were probably introduced by humans migrating from South America during pre-Columbian times. Another, *S. richteri* was introduced to North America from northern Argentina through Mobile, Alabama, in 1918, also probably in ballast soil.

The Imported Red Fire Ant, Solenopsis invicta

S. invicta is known as the imported red, or sometimes imperial, fire ant (it will be referred to in this study as the red fire ant). Adults are reddish to dark brown, and occur in five forms. Minor workers, usually less than 5 mm long, survive between 30 and 60 days. Major workers grow to just over 5 mm and survive between 60 and 90 days. Winged males, each about 10 mm long, survive up to 180 days. The winged queens, also about 10 mm long, survive between two and six years.

Red fire ant mounds vary in size, usually in direct proportion to the size of the colony. A mound 60 cm in diameter and 45 cm high, for example, may contain about 100,000 workers, several hundred winged adults, and one queen. Mounds can be found in all types of soils, except swamplands and dense forests. Mounds in clay soils are generally symmetrical and dome-shaped, while those in sandier soils tend to be shaped irregularly. Single-queen colonies can build between 70 and 250 mounds per hectare. With multi-queened colonies, as many as 1500 mounds per hectare have been observed.

Scientists calculate that, since their introduction in 1940, the total amount of living red fire ant biomass in North America has grown to about 2.5 million metric tons. From its original point of entry through the Gulf of Mexico, the North American red fire ant population currently inhabits 11 states stretching from Maryland to Texas. Its presence has recently been documented in Puerto Rico, and it continues to expand its range. Coupled with the harm it causes to human systems—particularly agriculture, housing development, and public safety—this rapid increase in the size and range of the North American red fire ant population classifies it as an invasive species.

Even experts can have difficulty distinguishing between the very aggressive red fire ant and less aggressive fire ant varieties—such as tropical and southern fire ants—in the field. Red fire ants swarm, bite, and sting when disturbed and, within 24 hours, a pustule-like sore forms at each sting site. These sores itch intensely, and scratching can rupture the skin and lead to secondary infection and scarring. A small proportion of people are highly allergic to fire ant stings, and require immediate medical attention. Be warned against unsupervised investigation, especially of any red fire ant mounds you find in fields.

Defining Native, Introduced, and Invasive Species

BACKGROUND

Humans have been introducing species from one continent to another, within particular continents, and among discrete ecosystems for many generations, both intentionally and unintentionally. To understand the science behind introduced species, students first need the skills to distinguish between native, introduced, and invasive species. In this activity, students will observe and identify native, introduced, and invasive species in familiar environments such as their school grounds and neighborhoods. Through their research, students will determine which of these species have been established for thousands of years, which were introduced more recently from other environments, and which of these later species have become invasive.

In this activity, students will make their first entries in their introduced species journals as they list species they observe during a walk around the familiar environments of their school grounds and neighborhoods. Back in the classroom, students will then use field guides and other resources to identify their listed species. Finally, students will create a local species chart, classifying the species they've listed and identified according to whether they are native, introduced, or invasive.

Upon seeing a white oak or silver maple, for example, students will include it on their list, identify it as native to North America, and incorporate it into their chart as a Native. Listing a tulip, a student will discover that it is native to Turkey, and chart it as Introduced. Students may list starlings, then learn that starlings, after being introduced in the 1890s, disrupted ecological processes by driving native birds out of their nests, threatening some bird species with extinction. Starlings continue to damage human systems by disrupting agriculture and spreading histoplasmosis, a fungal disease of the lungs. Students will therefore chart them as Invasive.

To get students started, photocopy and hand out the opposite page (page 6), which contains the "first chapter" in the ongoing red fire ant storyline. Have them use this page to begin their journals.

OBJECTIVE

To observe, identify, and classify native, introduced, and invasive species in familiar environments.

TIME MANAGEMENT

This activity may be completed in two class periods or more, depending on how much time you wish to allot for student observations and research. A minimum of one class period should be provided at the end of the activity for class discussion and/or guest presentations.

CONVERSION NOTE

Metric measurements are used throughout this book. Students may not be able to readily visualize measurements such as a square kilometer (km^2), square meter (m^2), or hectare (ha). For comparison, some conversions are provided below:

$1\ km^2 = 1{,}000{,}000\ m^2$, which is a square with sides 1,000 meters long

$1\ ha = 10{,}000\ m^2$, which is a square with sides 100 meters long

$100\ ha = 1\ km^2$

$1\ km^2 = 0.386$ square miles

$1\ ha = 2.477$ acres

PROCEDURE

MATERIALS

Field guides and other resources

Student journals (see page 10)

Art supplies

Camera (optional)

Guest speaker (optional)

CLASSROOM ORGANIZATION

Each student should keep their own introduced species journal. Depending on the classroom dynamic you wish to foster, you can lead the entire class around the school grounds and point out species for them to list. Or, you can divide the class into small groups and have each focus on a different section of the grounds. This way, student lists will contain a wider variety of species for identification and discussion.

TEACHING NOTE

Students may want to display their charts, sketches, and photographs. Depending on how many groups or classes are conducting this activity, you may want to organize a local species fair to allow individual, groups, or classes of students to present their observations, identifications, and species classifications to their peers and others.

1. Lead students on a walk around the school grounds or a local neighborhood, and have them list some of the organisms they observe. Common names will suffice, but students should describe, sketch, and perhaps photograph the species they include on their lists.

2. Depending on the amount of time you wish to allot, students may also list organisms they observe elsewhere in their local communities. Encourage students to list a wide variety of organisms, including insects, other invertebrates, and plants.

3. Back in the classroom, have students use field guides and other resources to identify the plants and animals they've listed and sketched. Once all listed organisms have been identified, have students construct a local species chart, such as Figure 1 on page 10, for inclusion in their journals.

4. Lead a class discussion for students to compare and contrast their findings about the native, introduced, and invasive species they've observed, identified, and charted. Encourage students to refer to their journals, and to note any questions or considerations they feel remain unanswered or wish to explore. The questions outlined below may be used to guide the discussion.

- Introduced species generally fail to become established in a new environment. What biological and physical characteristics would enable a species to become established upon being introduced? What other conditions would assist an introduced species in becoming established?
- What might happen when an introduced species comes into contact with native species? What advantages or disadvantages might native species have over introduced species? How might these increase or decrease a native species' chances of survival? An introduced species' chances?
- How do some introduced species become invasive in new environments? What are some examples of disruptive and harmful impacts of introduced species?
- What are some of the ways humans resist invasive species and try to control their disruptive and harmful impacts?

These questions outline topics which will be covered in detail in subsequent activities. Students can copy the questions in their journals, adding to their answers as they progress through the activities.

Generally, introduced species fail to become established due to damage from a lengthy journey, lack of food supply, and being killed by other organisms. Successful animal invaders are usually small, highly mobile, have high birth rates, are not picky eaters, and can adapt to a wide variety of ecological niches. Contact between introduced and native species—animal, plant, and microscopic—can have several possible results: competition, extinction, hybridization, and chemical change. Native species, having developed over centuries in an environment, are likely to be better adapted to that environment than new arrivals. However, introduced species, if they have few or no predators and plentiful nutrients, can disrupt established ecological processes and cause harm to human systems. If this occurs, they become known as an invasive species. Humans resist invaders using a broad range of means, such as closing or minimizing invasion pathways, instituting biological control measures, employing containment or eradication procedures, and establishing and enforcing national standards and policies.

SUGGESTIONS FOR FURTHER STUDY

Most introduced species don't get the chance to become invasive, and never become established in new environments. But those that do can disrupt ecological processes and harm human systems. Because students will be acquiring tools and skills to quantify the impacts of invasive species, they need to begin forming an idea of just what "disruption" and "harm" can mean. Begin with a brainstorming session; have students first identify some ecological processes, such as nutrient cycling, energy flowing through food webs, and species colonization and succession. Then have them draw on their existing knowledge to brainstorm how they think an invasive species might disrupt one or more of these processes. How has one of the invasive species they identified in this activity caused ecological disruption? How has a species with which they're familiar, but which they might not have identified in this activity, caused ecological disruption?

Students can also identify some human systems, such as economic systems involving agriculture and forestry, infrastructure systems involving roads, houses, and public works, and safety systems involving human and community health. How has an invasive species in this activity damaged a human system? Have students keep notes about their brainstorming session in their journals, and encourage them to revisit their notes as their understanding evolves.

Encourage students to distinguish between ecological processes that occur naturally, such as niche exploitation and range expansion, and ecological processes that are dramatically altered through human introduction of potentially invasive species.

FIGURE 1

After students have identified the species they included on their lists, have them construct a local species chart classifying them according to whether they are native, introduced, or invasive. They should include brief descriptions of the characteristics on which they based their decisions. The chart at right may be used as a model.

SPECIES CLASSIFICATION

	NATIVE	INTRODUCED	INVASIVE
Plant	silver maple *Acer saccharinum* evolved over centuries in eastern North American climate and soil	Japanese maple *Acer palmatum* originally from China, Korea, and Japan	eucalyptus *E. dalrympleanna* messy; bark and limbs peel and fall
Animal	bald eagle *Haliaestus leucocephalus* indigenous to North America	skylark *Alauda arvensis* native to Eurasia and Africa; introduced to Hawaii and British Columbia	European starling *Sturnus vulgaris* introduced from Eurasia in 1890s; disruptive and harmful
Plant	cattail *Typha latifolia* indigenous to North American wetlands and marshy areas	tulip *Tulip sp.* originally from Turkey; won't grow wild, so not invasive	wild carrot *Daucus carota* introduced as ornamental from Eurasia; a weed

Student journals

Having students keep their own introduced species journals during this study will provide them with a record of how their knowledge grows. Journals also aid with assessment. Encourage students to make daily entries in their journals. These may include definitions, thoughts about a class discussion, hypotheses on the outcome of an experiment, as well as data from that experiment, artwork, and photographs, either of their own or culled from various sources. Students should be encouraged to record questions directly in their journals, and to provide answers to those questions as their understanding evolves.

Writing helps students think through problems. It helps them formulate ideas, recognize what they already know, and highlight what they want and need to know. Read the journals on a regular basis, checking, initialing, or sampling them to mark the end of an entry. Because writing comments on every student's journal can be very time consuming, you might check off student completion of a particular journal entry in your grade book, and briefly refer to specific entries during the next class day as you return to your lesson plan. Or, you might cite one or two student entries or questions as a means of initiating class discussion. In this way, students will know their journals are being read, and that their questions are being responded to. This will encourage them to continue expanding and revisiting their journal entries.

You may wish to periodically assign a specific letter or number grade. If you check the entries regularly, assigning a specific grade to individual students can be accomplished quickly. Students might also periodically conduct a peer review. A good way to organize this might be to divide the class into small groups and have each group comment upon its members' journals. This provides a means not only for engaging students in each other's work, but also for demonstrating how constructive criticism may be used to address problems and make decisions.

You may wish to refer to the NSTA publication, *How to Write to Learn Science*, by Bob Tierney for more ideas on organizing student writing, journal-keeping, and related pedagogy.

WHAT ARE SOME RESOURCES IN YOUR COMMUNITY?

You may want to invite a guest speaker to answer student questions about what they've learned. Master gardeners, nursery personnel, birders, and other scientists and naturalists might accompany the class as it surveys local species. Representatives from Exotic Pest Plant Councils, which cooperate with Native Plant Societies and may be located through them, are also excellent resources. If your area is low in biodiversity, have experts bring in specimens illustrating the classifications of native, introduced, and invasive species. Or share your own research and experiences with students. The list on this page provides contact information for some native plant, botanical, and exotic pest plant societies in the U.S. and Canada. The New England Wildflower Society publishes an annual listing of clubs, societies, and councils in the U.S. and Canada. See the Resources section, beginning on page 56, for additional contacts.

SOUTHEAST

Alabama Wildflower Society
Rte. 2, Box 115
Northport, AL 35476

Arkansas Native Plant Society
Rte. 1, Box 282
Mena, AR 71953

Georgia Botanical Society
1676 Andover Ct.
Doraville, GA 30360

Florida Native Plant Society
PO Box 680008
Orlando, FL 32868

Louisiana Native Plant Society
Route 1, Box 151
Saline, LA 71070

North Carolina Wildflower Preservation
Totten Center, 475-A, UNC-CH
Chapel Hill, NC 27514

Mississippi Native Plant Society
202 North Andrews Avenue
Cleveland, MS 38732

Southern Appalachian Botanical Society
Newberry College, 2100 College St.
Newberry, SC 29108

Tennessee Native Plant Society
Dept. of Botany, UT-Knoxville
Knoxville, TN 37916

Virginia Native Plant Society
Box 844
Annandale, VA 22003

SOUTHWEST

Arizona Native Plant Society
PO Box 41206
Sun Station, Tucson, AZ 85717

California Native Plant Society
1722 Jade St. #17
Sacramento, CA 95814

Colorado Native Plant Society
Box 200
Fort Collins, CO 80522

El Paso Native Plant Society
7760 Maya Avenue
El Paso, TX 79912

Hawaiian Botanical Society
Botany Dept., University of Hawaii
3190 Maile Way
Honolulu, HI

Mojave Native Plant Society
8180 Placid Drive
Las Vegas, NV 89123

Northern Nevada Native Plant Society
Box 8965
Reno, NV 89507

Native Plant Society of New Mexico
Box 5917
Santa Fe, NM 87502

Oklahoma Native Plant Society
2435 South Peoria
Tulsa, OK 74114

Native Plant Society of Texas
Box 891
Georgetown, TX 78627

Utah Native Plant Society
Box 520041
Salt Lake City, UT 84152

NORTHEAST

Botanical Society of Washington
Dept. of Biology—NHB 166
Smithsonian Institution
Washington, DC 20560

Josselyn Botanical Society
Deering Hall, UM-Orono
Orono, ME 04469

Maryland Native Plant Society
14720 Claude Lane
Silver Spring, MD 20904

New England Wildflower Society
180 Hemenway Road
Framingham, MA 01701

New Jersey Native Plant Society
Box 1295R
Morristown, NJ 07960

New York Flora Association
New York State Museum, 3132 CEC
Albany, NY 12230

Pennsylvania Plant Society
316 4th Avenue
Pittsburgh, PA 15222

West Virginia Native Plant Society
Box 2755
Elkins, WV 26241

NORTHWEST

Alaska Native Plant Society
PO Box 14163
Anchorage, AK 99514

Idaho Native Plant Society
316 East Myrtle Street
Boise, ID 83706

Minnesota Native Plant Society
220 BioSci Center, UM-St. Paul
St. Paul, MN 55108

Montana Native Plant Society
Box 8783
Missoula, MT 59807-8782

Native Plant Society of Oregon
Box 902
Eugene, OR 97440

Washington Native Plant Society
Box 576
Woodinville, WA 98071-0576

Wyoming Native Plant Society
1604 Grand Avenue
Laramie, WY 82070

MIDWEST

Illinois Native Plant Society
20301 East 900 North Road
Westville, IL 61881

Indiana Native Plant Society
6106 Kingsley Drive
Indianapolis, IN 46220

Kansas Wildflower Society
17th & Jewell Streets
Topeka, KS 66621

Ohio Native Plant Society
6 Louise Drive
Chagrin Falls, OH 44022

Wildflower Association of Michigan
Box 80527
Lansing, MI 48908-0527

OTHER SOCIETIES

American Penstemon Society
1569 Highland Court
Lakewood, CO 80226

Chihuahuan Desert Research Institute
Box 1334
Alpine, TX 78739

Canadian Wildflower Society
4981 Highway 7 East, Unit 12A #228
Markham, Ontario, Canada L3R 1N1

Eastern Native Plant Alliance
Box 6101
McLean, VA 22106

National Council of State Garden Clubs
Box 860
Pocasset, MA 02559

Florida Exotic Pest Plant Council
3205 College Avenue
Ft. Lauderdale, FL 33314

Introduced Species Dispersal

OBJECTIVE

To investigate how introduced species disperse in new environments.

TIME MANAGEMENT

If you wish to involve students in the preparation and set up of this activity, Parts A and B may be completed over the course of one week. If you set up Part A beforehand, students can complete it in a double-period lab. Part B can be completed in one class period, with a portion at the end allotted for discussion.

BACKGROUND

In Activity 1, students observed, identified, and classified species in their local communities according to whether those species are native, introduced, or invasive. As students discussed their findings, perhaps with a guest speaker, they learned that the majority of introduced species fail to become established in a new environment. What characteristics, both biological and ecological, enable a species to become established once it's introduced? If an introduced species succeeds in becoming established in its new environment, what happens next? How, and when, does an introduced species become an invasive species?

In this activity, students acquire tools and skills for addressing these and other questions about introduced species dispersal. In Part A, students use crustaceans under laboratory conditions to investigate how an introduced species disperses in a new environment. In Part B, students use maps to investigate how environmental conditions can affect dispersal patterns, using the Africanized honeybee (*Apis scutellata*) as a model. In both parts, students explore the relationship between existing environmental conditions and behavioral adaptations in introduced species.

The level of success an introduced species has depends on a great many factors. Foremost is its physical ability to locate and exploit an ecological niche—to find and consume food. Other significant factors include the species' ability to physically adapt; the availability of appropriate nutrients and water; existing competitors and predators; and temperature and oxygen levels. These are the same sorts of environmental limitations that control the growth and spread of all populations.

If nutrient availability is high and competition and predation are low, for example, an introduced species has the potential to reproduce and spread over increasingly large geographic areas. Another significant factor is ecosystem stability. While the colonization and spread of introduced species is a complicated and much debated process, ecologists generally agree that the process is dynamic and usually involves some agent or agents of disturbance.

Students need to appreciate that the introduction of a new species, intentional or not, constitutes a disturbance within an established ecosystem. Further, a human-caused or natural disturbance can increase the likelihood that an opportunistic introduced species will be able to colonize and spread within a new environment.

For example, the lizard genus *Anolis* is distributed throughout northern South America, Central America, Mexico, the southeastern United States, and the West Indies. Within the West Indies, these lizards have undergone extensive adaptive radiation, have exhibited a wide variety of morphological types, and occupy a wide variety of ecological niches. In the Greater Antilles alone, *Anolis* species range in size from less than 35 mm long to lengths over 150 mm.

Scientists think these lizards are indigenous to South and Central America, and that they colonized the West Indies by traveling over water rather than by means of land bridges. Scientists think this because there is currently an abundance of diverse lizard species in the West Indies. Similar lizard species occur from one island to another, and they often occupy quite similar habitats. Such species are called ecomorphs, since they show great similarity in behavior, habitat, and morphology but are not closely related phylogenetically.

This activity uses pillbugs to provide an opportunity for students to explore many different influences on species colonization, dispersal, and range spread. Ecological succession, nutrient quality and population distribution, population growth rates, predator-prey interactions, and the effects of pollutants may all have an impact. Pillbugs make particularly convenient classroom tools for exploring such topics, as they are easily acquired, maintained, and manipulated.

PART A

SPECIES DISPERSAL

Part A is designed to allow students to investigate if and how a species disperses after being introduced to a new environment. Under laboratory conditions, students will introduce pillbugs to a new environment. Students will then measure how these introduced species disperse, by sampling and mapping them over two hours.

Involving students in construction and set up is desirable, but time will be saved if you complete the first steps yourself. Variations on this basic design are possible; some are described in this activity's Suggestions for Further Study (page 17). Depending on the time and resources available, students can explore additional variables to further illuminate what happens when a species is introduced to a new environment.

PROCEDURE

MATERIALS

One aquarium per student group (plastic containers are a less expensive alternative)

Forceps or soft paintbrush

Paper towels

Wood chips

Pillbug food: oatmeal, potato slices, lettuce leaves, and/or carrot peels

Plastic wrap (or a tightly-fitting aquarium lid)

Pillbugs (also called sowbugs)

Materials to create a micro-habitat: may include soil, dry sand, wet sand, petri dishes filled with water, small plants, heat lamp, shoe box, and rocks

Plaster of Paris (optional)

1. Pillbugs, also called sowbugs, are invertebrate crustaceans. Unlike lobsters, crabs, and crayfish, however, pillbugs live on land. Because they breathe through gills, pillbugs tend to inhabit moist, dark spaces. Invertebrates such as pillbugs are all around us, and perform many functions crucial to maintaining the balance of nature. For this activity, however, pillbugs will serve as an introduced species.

   In this activity, groups of four to five students will monitor a micro-habitat of their own. There are several options for creating micro-habitats. Possibilities include a 30-liter aquarium, deep plastic margarine tubs, or cut-off bottoms of large plastic milk containers. Aquaria can be lined with moistened paper towels; plastic containers can be filled with either paper towels or plaster of Paris. Plaster of Paris makes a uniformly damp bottom that holds test liquids well. If you choose plaster of Paris, mix it so that it is fairly stiff. If it's too thin, it will shrink from the sides when it dries and the pillbugs will crawl underneath.

   Pillbugs can be ordered from biological supply companies, or you can establish your own culture from pillbugs found under rocks and rotting logs. To store pillbugs, place several centimeters of soil in the bottom of a container. The soil should be from a wooded area with much organic matter. The container should contain wood chips, leaves, and stones. Sprinkle a little oatmeal on the surface of the soil, and add some potato slices and a few lettuce leaves or carrot peels. Place the culture where it won't be disturbed, making sure to keep it moist, and periodically add vegetable scraps.

2. Ask student groups to design their own micro-habitats in the containers provided. Their micro-habitats should be designed so that students can test how organisms disperse into a new environment. For example, students might explore how quickly the organisms disperse into a range of habitats by setting up aquariums that are partially lit and partially dark; another that has areas that are very damp and very dry; and another that has a pile of damp sand on one side and a pile of damp sawdust on the other. Try to test only one variable at a time.

3. Before proceeding, make sure that each group forms a hypothesis and prepares a data table to record the pillbugs' progress.

4. Just before the pillbugs are about to be placed into their micro-habitats, move them into dry petri dishes. Ten to 20 pillbugs per student group are ideal. Provide each group with pillbugs, and instruct students to gently place the pillbugs midway between the potential dispersal points. Placing the pillbugs on a small swatch of paper towel or cloth, and then placing the swatch in the habitat's center should suffice. Each student group should then cover their micro-habitat with plastic wrap so moisture is retained.

NOTE

To move pillbugs, gently use forceps or a soft paintbrush. Remind students to disturb the pillbugs as little as possible once they have been placed into the new environment.

5. Have students check their pillbugs' progress at regular intervals (every 10 to 15 minutes) to find out where the animals go and how quickly. If the micro-habitats are preconstructed, a double-period lab should provide sufficient time for students to gather migration data.

6. Have students record how many pillbugs were found in a different part of the micro-habitat, and at what intervals they were found there. Have students use their data to create a map of the pillbugs' dispersal pattern. (The Africanized Honeybee Dispersal Map on page 18 may be used as an example.) Students can create legends for their maps to completely describe environmental conditions, such as water temperature, substrate type, elevation changes, and so forth.

7. Have each student group give a brief presentation of their results to the class. What sort of environmental limitations confronted the pillbugs? What was the nature of the pillbugs' adaptations? Can students identify any pattern in the pillbugs' dispersal? Based on their data, do students think this introduced species can survive in its new environment? Groups should identify the test situations, any uncontrollable variables, any possible sources of error, and any unexpected results. Did the experiment prove or disprove the group's hypothesis?

8. The micro-habitats developed by the groups are probably not as healthy for pillbugs as the storage facility you built in step 1. Using a teaspoon, and being very gentle, students can move their pillbugs back to the petri dishes, then back to the storage facility at the end of the lab period. If the pillbugs were collected locally, you can return them to their approximate outdoor habitat after the activity.

PART B

MAPPING SPECIES DISPERSAL

Mapping is a fundamental tool scientists use to track how introduced species disperse and expand their ranges in new environments. In Part A, students constructed a dispersal map for their pillbugs. In Part B, they will expand on this basic mapping technique to gain an understanding of how an introduced species disperses over a larger geographic area.

As Part B proceeds, encourage students to keep in mind what they learned about dispersal in Part A. Part B is designed to help students understand how physiological and behavioral characteristics interact with environmental conditions to aid or inhibit dispersal. As students work through this procedure, encourage them to compare and contrast any differences and similarities they identify between Part A's crustacean introduced species and Part B's insect introduced species, the Africanized honeybee (*Apis scutellata*).

After seven years in the U.S., Africanized honeybees have dispersed throughout four states bordering Mexico. Isolated cases of bees arriving with cargo traffic have also been reported in other states. What factors have determined the direction and speed of their dispersal? In Part B, students examine a map to identify and discuss how geographic features and environmental conditions have affected the dispersal and range expansion of Africanized honeybees in the U.S.

PROCEDURE

MATERIALS

Africanized Honeybee Dispersal Map (page 18)

U.S. physical map

Other maps and reference materials

1. Photocopy and distribute the Africanized Honeybee Dispersal Map (page 18). It depicts the dispersal and range expansion of these bees in the U.S. Have students study the map's contents.

2. Have students use a U.S. physical map to identify the American Southwest's topography and major landforms. Have them compare its features with the expansion patterns depicted on the Dispersal Map. What physical features might have aided the bees' dispersal? What features might have inhibited it, or provided insurmountable barriers?

3. Have students examine other maps and reference materials showing climatic and environmental conditions—such as average annual temperature, average annual rainfall, and indigenous species—of the American Southwest. Invite students to consider the question: How might climatic and environmental conditions have aided or inhibited the bee's dispersal?

4. Have students construct a two-column chart for inclusion in their journals. Have them list the geographic, ecological, and environmental factors they identify in one column, juxtaposing a specific factor with its actual or hypothetical impact in another.

SUGGESTIONS FOR FURTHER STUDY

Using the procedure from Part A, students can run parallel experiments using identical environments but different animals, such as crickets; earwigs; protozoa typically found in mud or wet sand (*Euglena*, *Vorticella*, or *Blepharisma*); land snails; or land-phase salamanders.

If the pillbugs you used were locally collected, you may wish to have students monitor, over the course of a week, a local habitat which might contain pillbugs. Do not do this extension if your pillbugs were purchased. Students should mark off a test site of several square meters with string attached to stakes. Have them sample various places in this test site to estimate the environment's total pillbug population. If possible, have students take several samples to look at the stability of the pillbug population. The next week, release all the pillbugs used in the main activity into the center of the test habitat. Continue to monitor the site, counting how many pillbugs are located in its various parts. Students can draw a second dispersal map to compare dispersal under test conditions with dispersal under natural conditions.

How do red fire ants disperse?

Once introduced to Alabama, the original red fire ant population began a process of colonization and dispersal that continues today. Fire ants colonize and disperse by means of synchronized migration flights involving winged sexual females, or queens, and winged sexual males. On warm spring days after heavy rains, enormous flights of sometimes millions of individuals occur 100-250 meters in the air. After airborne fertilization, sexual females disperse anywhere from a few meters to well over a kilometer. From the air, they orient toward a favorable habitat, usually a partially vegetated, recently disturbed site.

Red fire ants also "hitchhike." The original invaders of Puerto Rico, for example, were probably carried by humans, as were red fire ants recently found in the western U.S. Humans also import red fire ants in the root balls of landscaping plants.

Once a queen lands, she breaks off her wings, digs a short tunnel, and seals herself in to rear a first brood of workers. If this mini-colony survives, the workers help produce the next, larger generation of workers, which produces another, and so forth. Eventually, a queen produces more queens, and the process begins anew.

Like most successful invaders, red fire ants possess physical and biological attributes that enable them to colonize new habitats easily. They are relatively small, highly mobile, and adaptable to a wide variety of ecological niches. They have high population growth rates and exhibit high fecundity. Despite widespread eradication and control efforts, aggressive, omnivorous red fire ants continue to colonize new habitats and disperse across a broad geographical range.

As students conduct Part A of this activity, have them record in their journals any comparisons they observe between the physiological and biological attributes of pillbugs and red fire ants, as well as between other introduced species with which they are familiar. What similarities and differences can students identify between how introduced species colonize new environments at both small and large scales? What organizing scientific principles hold true at both scales?

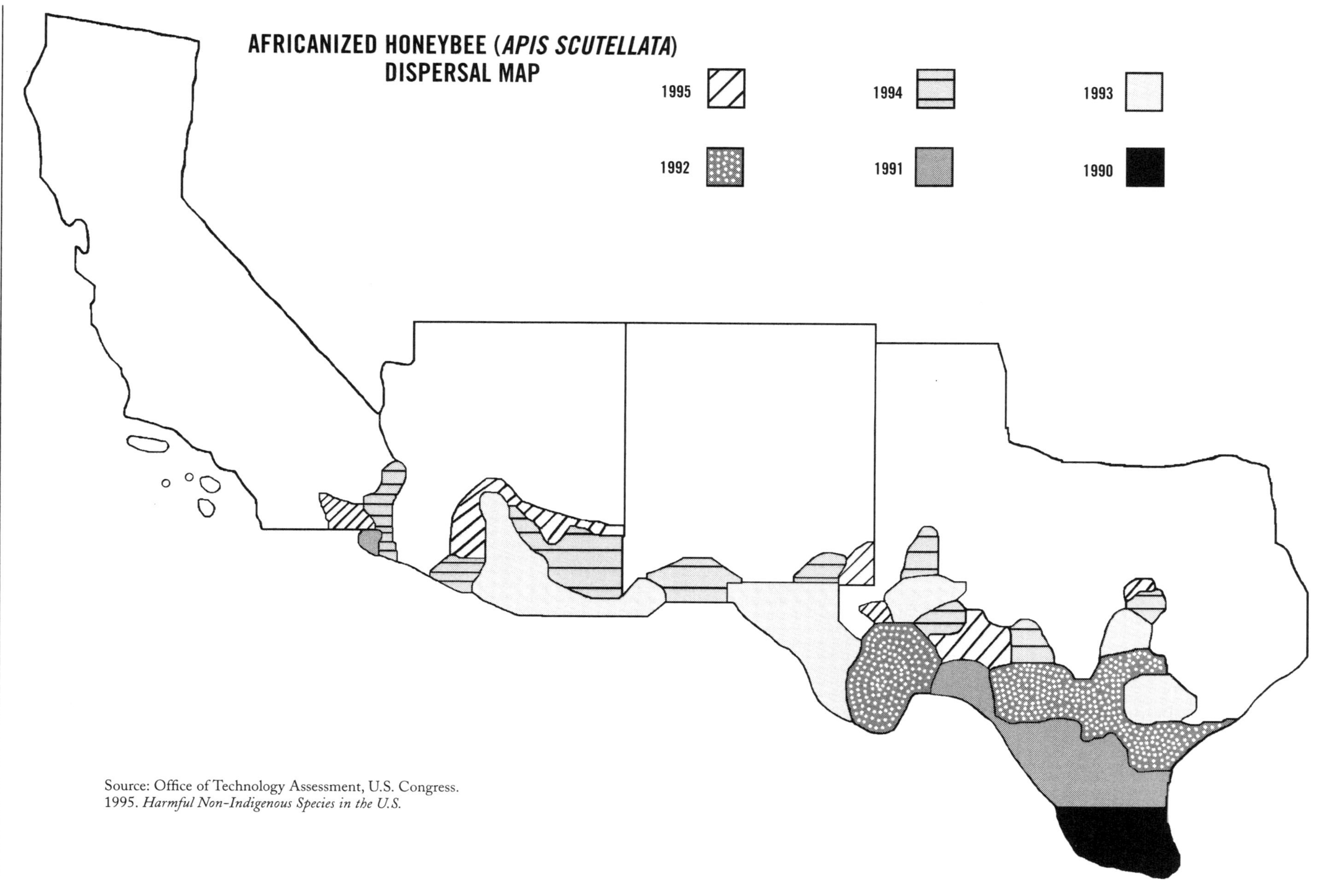

Source: Office of Technology Assessment, U.S. Congress. 1995. *Harmful Non-Indigenous Species in the U.S.*

Competing

BACKGROUND

Nineteenth-century diaries and letters of people who were seeing the American prairie for the first time paint a verbal portrait of a vista unlike anything in their experience. Many illustrate a monotonous sweep of waving grasses stretching to the horizon, containing few if any landmarks. Others depict the awesome challenge of venturing a forbidding expanse. Still others wonder if they'll ever make it through.

While the American prairie was certainly a formidable challenge to 19th-century travelers, it was also ecologically diverse and complex. Many of its features are apparent only upon close examination at a small scale. The monotonous, forbidding sweep some of those diaries and letters describe was actually comprised of a hundred different grass species and sheltered thousands of insect species. Numerous small mammalian herbivores—prairie dogs, jackrabbits, and ground squirrels—as well as their predators—coyotes and black-footed ferrets—hid in the grass. In some prairies, large American bison, pronghorn antelope, and elk dwelt side by side with cougars and red and gray wolves. Hundreds of bird species inhabited the prairie—meadowlarks, red-winged blackbirds, sedge wrens, sparrows, dickcissels, and the remarkable prairie chicken, whose mating rituals came to epitomize the prairie. Dozens of reptile and amphibian species inhabited the sweeping grasses. The cornucopia of animal life on the prairie is not always obvious because so many species live underground, and long grasses hide the aboveground species.

Though all prairies look superficially like seas of grass, each also contains a broad but specific variety of plant species. There are many types of prairie, broadly divided into tallgrass prairie, shortgrass prairie, mixed prairie, and high prairies bounded by mountains, like those between the Rocky Mountains and the Sierra Nevada–Cascade ranges. The species of animals and plants may differ enormously among these geographically separate prairies. Their dominant grasses may not even be the same.

Prairie rainfall is typically between 25 and 75 cm per year, but there is tremendous variation, with high evaporation rates exaggerating the

OBJECTIVE

To acquire tools and skills for understanding how an invasive species competes with native species.

TIME MANAGEMENT

This activity can be completed in two periods, a little over a week apart. Students will need a few minutes during interim classes to record plant data.

CLASSROOM ORGANIZATION

Students will be divided in groups for this activity.

effects of occasional severe droughts. In many prairies, the temperature fluctuates between bitter winter cold and extreme summer heat, with the effects exacerbated by the dearth of shelter and shade and by howling winds. Most prairies are typified by frequent fires, many started by lightning. Such fires may burn for many miles and eliminate tree seedlings and saplings that might otherwise dominate the prairie landscape if allowed to grow.

Since the first non-Native Americans traversed the prairie, increased human expansion into prairie habitat—and the ensuing development of that habitat for various purposes—has significantly altered what is now important agricultural land in the United States.

Prairie grasses are of two types: bunchgrasses, or tussock grasses; and turfgrasses, or rhizomatous grasses. Bunchgrasses, which originally dominated most American prairies, typically form clumps, or tussocks. All shoots grow straight up, and spread at their base is by sprouts. Turfgrasses form thick mats and reproduce and spread by underground stolons and rhizomes.

Bunchgrasses are generally much more liable to be damaged by intensive trampling and feeding by large, congregating herbivores. The introduction of livestock from Europe (especially cattle and sheep) has devastated many prairies. (In the Great Plains, bison had been present, but they did not graze in the same spots year after year. In the intermountain West, there were few or no bison.)

*Cheatgrass (*Bromus tectorum*), shown below, was introduced to the American West in the 1880s. Within 50 years, cheatgrass infested more than 200,000 square kilometers. Today it may be found in all 50 states.*

Eurasian turfgrasses, however, are well adapted to large congregating herbivores. Their low height and underground stolons are more resistant to trampling and grazing. Some Eurasian grass species were introduced in the Midwest and West deliberately, some accidentally. Many, like cheatgrass, have become invasive. Cheatgrass (*Bromus tectorum*) now occurs in all 50 states and is an exotic pest plant in the Midwest and West, infesting millions of hectares and, in some areas, lowering rangeland productivity by more than 50 percent. Cheatgrass burns easily and recovers far more quickly than the native plants it is replacing, so its presence increases the frequency and consequences of fires. Cheatgrass also does not support the hundreds of native insect species that rely on native bunchgrasses, and its dense growth does not favor small native herbs and shrubs that normally grow between the tussocks of native grasses.

Scientists think cheatgrass arrived in the inter-mountain West in the 1880s in impure seed, as did many weeds before seed purity laws were introduced at the turn of the century. The earliest records of cheatgrass come from wheat-growing areas. Through the first decade of the 20th century, cheatgrass was frequently seen along railroad rights-of-way. Within 50 years of its arrival, cheatgrass infested more than 200,000 square kilometers. Continuing problems with seed purity, plus the fact that cheatgrass seeds are easily carried on fur and remain viable in dung, helped it complete its spread in the inter-mountain West by 1930. The expansion of cheatgrass continues today despite efforts to contain it.

How do red fire ants compete?

Imported red fire ants have eliminated both species of native fire ants and the exotic black fire ant from suitable habitat throughout most of their current range. They have done this partly by resource competition—by eating food other fire ants would have eaten. Most ants are omnivores, feeding on any available animal or plant matter. Red fire ants have particularly broad tastes, preying on invertebrates, vertebrates, and plants. They are also aggressive scavengers of dead material.

Red fire ants also engage in interference competition, attacking and killing workers and queens of other ant species when they find them. The red fire ant out-competes a number of other ant species at food sources, including another exotic invader, the Argentine ant, by attacking them in great numbers. Even species that emit repellent defensive chemicals are eventually overwhelmed by the huge number of workers the red fire ant can mobilize on a food source. Some ant species remain in a region after the red fire ant has invaded, but largely in habitat undisturbed by fire or by humans, and often in reduced numbers. It's been suggested—but not proved—that some vertebrate predators, like lizards, are declining in numbers because red fire ants are out-competing them for food.

Red fire ants prey on many species of insects—ants, butterflies, moths, crickets, flies, beetles, etc.—plus small invertebrates like worms, ticks, and isopods. It is rarely established that this predation has an impact on the prey population, but the huge numbers of red fire ants present in each colony probably depress prey populations in some places. There has so far been little long-term, controlled, and replicated experimental research of the sort that would be required to detect a population impact.

Red fire ants may also be subtly but inexorably changing the nature of natural plant communities. They eat seeds; they prefer some species over others; and they move seeds about. However, their impact on plants has been studied only where they are agricultural pests.

PROCEDURE

MATERIALS

Seeds
- grass
- corn
- radish
- kidney bean
- nasturtium

Paper cups

Potting soil

Plastic bags

Student Worksheets (page 25)

Field guides

Cheatgrass specimen (optional)

1. Explain what cheatgrass is, and when it was introduced. If time allows, have students locate pictures of both undisturbed prairie and cheatgrass-infested prairies. (Library resources include encyclopedias and ecology books. Page 59 provides some Internet resources.) At first glance, both ecosystems might appear the same. But a closer look reveals that the ecologies of the two systems are quite different. Encourage students to explain what some differences are. For example, are tussock grasses still present? Are there different insects?

2. Invite students to consider possible explanations for the competitive advantage cheatgrass experiences in prairie ecosystems. Have students discuss these causes in groups, lead a class discussion, or have students speculate about the causes in a homework assignment. Have them record their conclusion in their journals, and then reconsider their conclusions after completing the procedure. Questions might include:

   - How does cheatgrass differ from the original native species?
   - Are there differences in root structure?
   - Is cheatgrass more drought resistant, insect resistant, or heat resistant than the original species?
   - Are there differences in reproductive patterns, seed structure, seed germination, seed viability, seedling size, or time to maturity?
   - How do different seed germination patterns and growth patterns affect the survival of various plant species?

3. In this experiment, the class will work with a variety of seeds to try to find possible answers to these questions. Because there are 15 different experimental and control groups of plants, assign each student group two or three seed combinations. The student groups will then set up, grow, and record data for their assigned seed combinations.

4. The following seeds are used because they vary in cotyledon number and size, thickness of seed coat, size, and length of germination time.

   grass: small seed, monocot, mid-length germination (5-6 days).
   corn: large seed, monocot, longer germination (6-7 days).
   radish: small seed, dicot, extremely short germination (2-3 days).
   kidney bean: large seed, dicot, mid-length germination (5-6 days).
   nasturtium: medium sized seed, dicot, extremely tough seed coat, long germination (7+ days).

5. Seeds can be planted in paper cups or small milk cartons which have been filled with well-moistened potting soil (mixture of equal parts perlite, vermiculite, and sphagnum peat). Drainage holes should be poked in the bottom of the container with a pencil. Ten seeds should be planted in each container according to the following plan:

   Container 1: 10 grass seeds
   Container 2: 10 corn seeds
   Container 3: 10 radish seeds
   Container 4: 10 bean seeds
   Container 5: 10 nasturtium seeds
   Container 6: 5 corn seeds, 5 grass seeds
   Container 7: 5 radish seeds, 5 grass seeds
   Container 8: 6 radish seeds, 5 corn seeds
   Container 9: 5 bean seeds, 5 grass seeds
   Container 10: 5 bean seeds, 5 corn seeds
   Container 11: 5 bean seeds, 5 radish seeds
   Container 12: 5 nasturtium seeds, 5 grass seeds
   Container 13: 5 nasturtium seeds, 5 corn seeds
   Container 14: 5 nasturtium seeds, 5 radish seeds
   Container 15: 5 nasturtium seeds, 5 bean seeds

6. Each planted container should be placed in a clear plastic bag to maintain humidity and prevent the soil from drying out.

7. Each student group will be assigned several (3-4) experimental containers. It is suggested that each group have at least a single species container and the remaining containers be mixed. Because data will be pooled at the conclusion of the experiment, this is an excellent opportunity to discuss the role of a large sample size in yielding reliable data.

8. The class should discuss what data they should record and how they will record their data. Groups should design a data chart that will organize their data and allow them to collect data efficiently and quickly. Or use the student worksheet on page 25. The following are suggested pieces of information that you may wish students to collect: seeds size, shape, mass, color, cotyledon number, thickness of coat, ability to absorb water, day or night temperatures at which the experiment was conducted, seedling height, development of first true leaves, stem diameter, and health.

9. After the groups have planted their seeds, encourage them to make predictions about what will happen in each of their containers. Have them record their predictions in their journals. When the seeds begin to sprout, encourage the groups to share their findings. Keep probing the students with questions. Which seeds are the first to sprout? What is the selective advantage to a species that has a short sprouting time? What are the disadvantages? Does this selective advantage depend on the specific climate in which the plant is growing? For example, what disadvantage might be experienced by an early germinator in a northern climate? Do all 10 of the seeds survive in each of the single species populations? What differences do you notice among the 10 seedlings in the single species populations?

10. Have student groups pool their results. Provide each student with a copy of the combined data. This experiment represents an excellent opportunity to have a real-life demonstration of the variability that exists among all members of a species. Where do the niches of the mixed populations overlap? How does this cause a problem to the survival of each of the species? If you were to introduce an aphid, a grasshopper, a snail, or some other plant-eating insect into your containers of growing plants, what effects might there be?

Kudzu (below) is a well known example of an introduced species that has altered environmental conditions in parts of America, especially the Southeast. It was introduced there from Asia in the late 1940s in an attempt to halt soil erosion. But environmental conditions which had traditionally held its growth in check—such as herbivores, nutrient availability, and range limitations—did not exist in the American Southeast. Its continued growth there has had a dramatic impact on other species, and kudzu currently overwhelms many hectares of Southeastern American landscape.

CULMINATING DISCUSSION

When all the data have been pooled, ask the class if any of the data could suggest some reasons why cheatgrass has survived in the prairie and the native species have been replaced. What other experiments could they suggest? If you could take a field trip to the prairie, what data would you gather that would help you to determine why cheatgrass has been so invasive? Ultimately, you will need to help students make the connection between their containers of growing seeds and the invasive cheatgrass population. For example, students may find that rapid germinators such as radish can interfere with the growth of seeds that take longer to germinate. You may wish to bring the discussion around to the idea that cheatgrass burns easily and is able to recover much more easily than the indigenous species of grass from the effects of fire. Abundant cheatgrass will increase brushfire frequency, therefore changing the ecosystem's overall temperature, which in turn changes the makeup of the ecosystem's inhabitant species. Moreover, cheatgrass' ability to recover rapidly from disturbance may prevent some indigenous grasses from growing. This competitive exclusion may be similar to the way rapidly germinating seeds interfere with the growth of seeds that have a longer period of germination.

STUDENT WORKSHEET

STANDARD INFORMATION ABOUT THE SEED ITSELF

(Average size, shape, mass, color, cotyledon number, thickness of coat, ability to absorb water, and temperature at which the experiment is conducted.)

DATE OF OBSERVATION

FIRST GERMINATION TIME

(Number of seeds sprouted.)

HEIGHT OF SEEDLING

LENGTH OF TIME TO DEVELOP FIRST TRUE LEAVES

STEM DIAMETER

GENERAL COLOR, APPEARANCE, HEALTH OF PLANT

(*i.e.*, a long "leggy" plant is tall, but not beccessarily healthy because its stems are weak.)

Hybridizing

OBJECTIVE

To acquire tools and skills for understanding how selection pressures can be used to control allele frequency in a population.

TIME MANAGEMENT

This activity may be completed in one class period.

BACKGROUND

One tool for students to acquire to understand introduced species comes from population genetics, which studies the factors involved in maintaining and changing the genetic constitution of populations, as well as the bearing such factors have on evolutionary processes like hybridization. Your students may have studied phenotypes, genotypes, alleles, and frequencies in their biology coursework, but it might be useful for them to refresh their memories, especially regarding the vocabulary of genetics.

A fundamental principle of population genetics is the Hardy-Weinberg Law. Mathematician G. Hardy and physician W. Weinberg demonstrated that, when mating is random and no factor favors one allele over others at a locus on a chromosome, the allele frequency in successive generations remains in equlibrium. (An allele is an alternative form of a gene at a particular site, or locus, on a chromosome.) Studies involving introduced species can use this law to predict the frequencies of genotypes, and so provide a basis for understanding influences on behavior derived from actual observations.

In this two-part activity, students examine how humans can manipulate gene frequencies within a population. Scientists, for example, are actively hybridizing Africanized honeybee populations to reduce their aggressive behavior. One way they are doing this is by "requeening" Africanized honeybee populations with non-aggressive queens, so that subsequent generations might exhibit less aggressive behavior.

In the 1950s, Brazilian researchers imported honeybees from Africa and developed an African-European hybrid, the Africanized honeybee (*Apis scutellata*). Some hybrid swarms escaped and further hybridized with domestic honeybees. These hybrid honeybees swarm more often, defend their nests more vigorously and effectively, and devote more of their resources (such as honey) to increasing their numbers rather than storing for shortages. They were first spotted in the U.S. in Texas in 1990, and have since spread their range among four states bordering Mexico (refer to the Africanized Honeybee Dispersal Map on page 18).

PART A
MODELING POPULATION GENETICS

Have students start with the premise that aggressive and non-aggressive behaviors in Africanized honeybees are determined by their genetic makeup, and that one gene locus with two alleles controls their level of aggression. (It's important to explain to students that this is a much-simplified model used only for the purposes of this experiment. Animal behavior is complex, and rarely determined by a single gene.)

If the allele for aggression is labelled "A" and the allele for non-aggression is labelled "a," individual bees can have one of three possible allele combinations: "AA," "Aa," and "aa." The combination of an individual bee's genes (its genotype) affects the trait we observe (its phenotype). "AA" individuals are aggressive while "aa" individuals are non-aggressive. An "Aa" individual, therefore, is also aggressive because its "A" allele dominates its "a" allele.

Starting with a population comprised of an "AA" parent and an "aa" parent, all the individuals in the first generation must have the "Aa" genotype, making them aggressive. In Part A of this activity, students calculate the frequencies of the appearance of the aggression genotype in subsequent generations. Have them assume that each individual is equally likely to contribute either of its two alleles to the following generation, and that matings between each pair of genotypes occur with a frequency corresponding to the product of the frequencies of those genotypes within the population.

This means that, in the second generation, all matings will be between "Aa" individuals. One quarter of those offspring will be "AA" individuals, half will be "Aa" individuals, and another quarter will be "aa" individuals. According to Hardy-Weinberg, this 25/50/25 ratio will remain in equilibrium throughout subsequent generations. By multiplying out the frequencies of possible matings in those subsequent generations, as in (0.25AA + 0.5Aa + 0.25aa)(0.25AA + 0.5Aa + 0.25aa), students will verify the Hardy-Weinberg equilibrium. (It should be noted that the Hardy-Weinberg Law's assumptions are never perfectly fulfilled in natural populations, and so must be considered wholly theoretical.)

PROCEDURE

MATERIALS

100 candies, 50 of one color and 50 of another

Paper bag

Paper cups

Student Data Sheet (page 31)

Specimen (optional)

1. In front of the class, set up the "genetic makeup" of your model bee population, and explain to students what you are doing. Let 50 candies of one color represent the dominant allele for aggression (A), and 50 candies of the other color represent the recessive allele for non-aggression (a). Place the 100 candies, 50 of each color, in the paper bag and shake to mix. Label one paper cup "Agg." for the aggressive phenotype, and the other cup "Non-agg." for the non-aggressive phenotype.

2. Randomly select two candies from the bag, and have students identify the phenotype they represent as aggressive or non-aggressive. Place each two-candy genotype in the appropriate phenotype cup. Have students record their identification on their Data Sheets by marking the appropriate column (Agg. or Non-agg.). Once the bag is empty, have students identify the genotype represented by the candies (AA, Aa, or aa), and mark the appropriate column on their Data Sheets.

3. Before continuing with further generations, have students use their data to calculate frequencies for phenotypes, genotypes, and alleles for generation 1. The equations below may be used as a guide.

$$\text{Phenotype} = \frac{\text{\# of aggressive individuals}}{\text{total \# of individuals in population}}$$

$$\text{Genotype} = \frac{\text{\# of individuals of particular genotype}}{\text{total \# of individuals in population}}$$

$$\text{Allele} = \frac{\text{\# of particular allele}}{\text{total alleles in population}}$$

4. Place all the candies from both cups back in the bag, and repeat this process for four more generations. Have students record the results for each generation, and perform their calculations.

What happens when red fire ants hybridize?

Scientists haven't found evidence that the red fire ant has hybridized with either of North America's two native fire ant species. But colonies have been found containing individuals which are apparently hybrids of the red and black fire ant (*S. richteri*), introduced in 1918 through Mobile, Alabama. The concern is that these hybrids may be able to extend the red fire ant's range northward. This is due to the well-known phenomenon of hybrid vigor: hybrids between two closely related species—or two genetically distinct populations within a species—become more vigorous. They may, for example, grow to be larger, or they may tolerate climatic conditions that would otherwise be too extreme for either parental strain. The distribution and spread of hybrid fire ants is being actively studied by agricultural researchers.

PART B

SIMULATING SELECTION PRESSURES

One strategy scientists use to address the invasion of Africanized bees is to requeen bee populations with non-aggressive queens. The idea is to eventually lower, maybe even eliminate, the number of aggressive individuals in the overall population. In the following model, students examine what happens when mating is no longer completely random, as it was in the Part A model. In Part B, the "bees" are only allowed to mate with a "queen" with a non-aggressive genotype (aa).

PROCEDURE

1. Explain to students that one paper cup is labelled "Agg." for the aggressive phenotype, another paper cup "Non-agg." for the non-aggressive phenotype, and a third paper cup "Queen" for the recessive, non-aggressive allele. Let 25 candies of one color represent the dominant allele for aggression (A), and 75 candies of the other color represent the recessive, non-aggression allele (a). Place 50 of the 75 candies in the "Queen" cup. Place the remaining 25 "a" candies and the 25 "A" candies in the paper bag, and shake.

2. In front of the class, select one candy from the bag and one from the "Queen" cup; this pair represents a genotype. Have students record the appearances of aggressive and non-aggressive individuals on their Data Sheets as they occur. Place the paired candies in the appropriate cup (Agg. or Non-agg), assuming that aggression is controlled by the dominant allele (A) and non-aggression by the recessive allele (a).

3. Have students determine phenotype, genotype, and allele frequencies using the same method as in Part A. Place the candies from the cups back in the bag, and repeat this process for four more generations. Have students record their data and perform their calculations after each generation.

4. Repeat this process, but this time don't allow all the "bees" with an AA or Aa genotype to mate. In other words, don't place all the aggressive pairs back in the bag after the second generation. In each subsequent generation, remove a few more AA and Aa pairs.

5. Once students have completed their calculations for selecting against aggression, ask them what has happened to phenotype, genotype and allele frequencies since mating is no longer completely random.

MATERIALS

100 candies, 25 of one color and 75 of another

Paper bag

Paper cups

Student Data Sheet (page 31)

Graph paper (optional)

Specimen (optional)

SUGGESTIONS FOR FURTHER STUDY

You may want to have students plot the appearance frequencies of both "A" and "a" alleles on a sheet of graph paper, plotting frequencies on the y-axis and generations on the x-axis. This will provide a visual reference for students to see the relationship between the appearances of the aggressive and non-aggressive alleles.

For several decades, honeybees in the U.S. have been dying. The number of managed colonies fell from about six million in 1940 to three million in 1996. What has been the impact on U.S. agriculture? Students can research some possible causes and, in a class discussion, evaluate the impact of Africanized bees on U.S. honeybees. The sidebar below, "Pollination and U.S. Agriculture," may be used as a starting point for student research.

RESOURCE

The Bee Keepers homepage: http://ourworld.compuserve.com/homepages/Beekeeping/right.htm

Students may want to research in more detail the methods by which beekeepers introduced European honeybees into Africanized swarms. How are bee colonies captured in the wild? What methods do beekeepers use to control the behavior of this potentially unwieldy, yet extremely useful, insect species? Have students conduct on-line or library research to engage these and other questions. Even though students are conducting research indoors, they should be reminded never to approach or tamper with a beehive or bee colony in the wild, as the behavior of its inhabitants is unpredictable and potentially dangerous.

Science fiction is frequently based on "what ifs." Students might investigate the genre of science fiction movies that capitalized on the public's fascination with genetically-engineered species. When did these movies first begin appearing? Which ones were based on actual situations, processes, or technologies? Can students identify recently released movies, books, commercials, or television shows that appeal to the public's curiousity about genetic engineering?

Pollination and U.S. Agriculture

Honeybees (*Apis mellifera*) pollinate up to $10 billion worth of apples, almonds, and other crops every year. Their value as pollinators exceeds the $250 million annual honey production. Many fruit growers lease hives from beekeepers, who move their bees north for the summer and south for the winter.

Pollination is crucial for the production of viable seeds and for fully-developed, tasty fruit. To find out how well an apple was pollinated, simply slice the fruit crosswise and check for two seeds in each of the five seed pockets. Count only fully-developed, not withered, seeds. Some supermarket apples will only have two or three seeds, while large gourmet apples may have six or eight.

A few entomologists are hoping the decline in honeybee populations will give other native bees an opportunity to grow their populations and expand their ranges. Native bees, in some cases, are not as vulnerable to diseases and pests as the introduced European honeybee. Bumble, squash, and gourd bees have all been found in fields in which honeybees were once the primary pollinators. The efficiency of these other bees in their pollination possibilities for agriculture is, however, unclear. Very little is known about the sizes and distribution of the 3,500 or more species of native bee populations. Native bee populations are more abundant in the American Northeast than in the Midwest, where vast expanses of monoculture crops have made them scarce. Some roadsides and wastelands have been dedicated to prairie wildflowers in an effort to encourage native bees to return.

STUDENT DATA SHEET

PART A

MODELING POPULATION GENETICS IN SUCCESSIVE GENERATIONS

generation	phenotypes		genotypes			alleles	
	Agg.	Non-agg.	AA	Aa	aa	A	a
1.							
2.							
3.							
4.							
5.							

PART B

SIMULATING SELECTION PRESSURES

generation	phenotypes		genotypes			alleles	
	Agg.	Non-agg.	AA	Aa	aa	A	a
1.							
2.							
3.							
4.							
5.							

PART B, STEP 4

SELECTING AGAINST AGGRESSION

generation	phenotypes		genotypes			alleles	
	Agg.	Non-agg.	AA	Aa	aa	A	a
1.							
2.							
3.							
4.							
5.							

Rapid Evolution and the House Sparrow

OBJECTIVE

To examine and analyze evidence of evolutionary change in the descendants of the pioneer population of introduced house sparrows in North America.

TIME MANAGEMENT

This activity may be completed in one class period.

BACKGROUND

Fifty house sparrows (*Passer domesticus*) were successfully introduced from Europe to North America through Brooklyn, New York, in 1853. The species spread westward reaching Death Valley, California, after 60 years; Vancouver, British Columbia, after 70 years; and Mexico City in about 1930. House sparrows were also introduced to the Hawaiian Islands from New Zealand in 1870, where they had previously been introduced from England. By the early 20th century, *P. domesticus* had dispersed across Canada, the United States, and Mexico.

By the 1930s, ornithologists had begun noticing that *P. domesticus* populations were exhibiting distinct racial differences, such as plumage color, weight, bill length, and variations in bone size. This suggested that widely dispersed populations of *P. domesticus* were becoming genetically distinct within the different geographic regions they now inhabited. Scientists concluded that geographically disparate populations of species *P. domesticus* were evolving into distinct races.

Traditional scientific estimates of the minimum time required for racial evolution in birds had been longer than 4,000 years. These estimates were based on scientific studies of European bird populations. In the 1960s, biologists in the United States hypothesized that North American *P. domesticus* populations had produced adaptive evolutionary responses to their new environments in as little as 50 years.

This activity is designed to help students understand how evolutionary adaptation in introduced species occurs, as well as what forms it can take. Students will analyze evidence derived from the 1960s study, and acquire tools for understanding evolutionary adaptation among other introduced species. Two such tools are Bergmann's and Allen's Rules of Ecogeography, respectively (see "How does geographic location influence species size and shape?" on page 36); another is an understanding of how variations relate to climatic conditions. Encourage students to apply these tools to species they have studied or with which they are familiar, and to incorporate their insights in their journals.

PROCEDURE

MATERIALS

Figure 1 (page 36)

Figure 2 (page 36)

Figure 3 (page 37)

1. Tell students that, in one major study, scientists collected skeletons of 1,825 adult house sparrows from 30 locations in the U.S., Canada, and Mexico. Using a dial caliper accurate to 0.05 mm, measurements were made of the skull bones, the pectoral girdle and wing bones, and the leg bones. The average (mean) dimensions of these bone measurements were arranged mathematically to represent the overall sizes of the birds and, using these individual measurements, the scientists assigned overall size ranges for birds among the 30 locations. The smallest skeletons were described as Size 1; the largest as Size 10. The interval between each size category equals about 1/3 of one standard deviation (see page 34).

2. Having photocopied pages 36 and 37, distribute them to individual or groups of students. Have students review the materials they contain, and then point out to them that Figure 1 graphically depicts the relationship between the geographic distribution and the skeletal sizes of the birds measured in the study. Figure 1's numbers denote skeletal sizes, and its contour lines denote extended geographic ranges. Have students discuss what this illustration signifies. (You may also want to have them refer to a physical atlas or classroom map.) Point out that these data were collected in the 1960s, only 100 years after 50 sparrows were released in Brooklyn, New York.

3. Have students examine Figure 2, which illustrates the mean weight of sparrows plotted against isophanes. Isophanes are values calculated from latitude, longitude, and altitude which, in combination, reflect overall climatic conditions. Have students note the mean weight and isophane values of two sparrows, such as one from Oakland, California, (27.3 g at 35° N) and one from Salt Lake City, Utah (29 g at 49° N). Then have students study the line of the graph, which shows the trend of weights against isophanes at 14 sites in North America. Invite students to use both Bergmann's and Allen's Rules of Ecogeography to make connections between the data in Figure 1 and Figure 2.

4. Figure 3 depicts plumage variations among sparrows from five U.S. regions: the Pacific Northwest (and Vancouver, BC), the Southwest, the Midwest, the East Coast, and Hawaii. Have students discuss the variations they observe, and consider how climate conditions among these regions may have influenced plumage differences. Lead a class discussion, and encourage students to make notes in their journals as it progresses. Questions to guide your discussion can include:

Q: What interpretation can be made of the generalized contour diagram of geographic distribution in mean values of house sparrow skeletal measures in North America?
A: In general, birds with the largest bones are concentrated in north central North America. Smallest skeletons are found in southern regions or California's hot valleys.

Q: How does Bergmann's Rule apply to these data?
A: Bigger birds were found in colder ranges.

Q: In conformance with Allen's Rule, what might one expect to find in house sparrow skeletons in North America?
A: Sparrows with the smallest bodies and longest legs live in hot ranges.

Q: How do these data suggest the house sparrow's evolutionary adaptation in North America responded to isophanes?
A: The higher the isophane, the heavier the birds.

Q: How do these data illustrate Bergmann's Rule?
A: Higher isophanes tend to occur in colder zones.

Q: What environmental factors appear to have influenced plumage variations in the house sparrows depicted in Figure 3?
A: Answers will probably vary, but should involve a discussion of climatic factors and environmental conditions. Possibilities can include: regional temperature and humidity variations; variations in intensity and duration of sunlight; adapting to the need to protect themselves from predators; developing their own predatory habits; and mating.

What is standard deviation?

Standard deviation is a measure of how far away the points in a data set are from the average. Standard deviation measures dispersion—how spread out the points in a data set are from the average. A data set's standard deviation is often reported as an absolute error, using the same units as the measurement.

Standard deviation is equal to the square root of the mean of the squares of the deviations from the arithmetic mean of the distribution. To find average (mean) use the equation at top right; to find standard deviation (where n = the total number of data points) use the bottom right equation:

$$X = \text{Average} = \frac{\Sigma \overline{X}}{n}$$

$$\delta = \text{Standard Deviation} = \sqrt{\frac{\Sigma(\overline{X} - X)^2}{n - 1}}$$

SUGGESTIONS FOR FURTHER STUDY

Have students research how systematists use genetic information to determine the evolutionary relationships among species. What is molecular phylogenetics? Have students look up these and other key words and report on their findings. The Internet is a starting point, but library research—especially using the *Reader's Guide to Periodic Literature*—is most desirable.

Have students investigate how the introduced species, the brown tree snake (*Boiga irregularis*), might have evolved from its original biological condition in Australia to its current biological condition on the island of Guam. (This will set the stage for Activity 7, Balancing Human Systems with Natural Processes.) Based on what they've learned in this activity, can students hypothesize about future evolutionary adaptations that might be experienced by this snake? The brown tree snake has recently been sighted on Oahu, Hawaii. How might environmental conditions there influence the evolutionary adaptation of *B. irregularis*? Students can reflect on these questions in their journals, and refer to them when they begin the culminating activity.

This activity has focused on terrestrial habitats. To broaden students' understanding, invite them to consider how evolutionary adaptation might occur among species in non-terrestrial environments. Students are often fascinated by oceans, and might be encouraged to apply their interest to this particular subject. Encourage students to also explore evolutionary adaptation among microscopic introduced species. Various bacteria and viruses—such as those that cause human discomfort, illness, and even death—are transported around the world by a variety of means. Students can research how new viral strains are affecting global health care. The United Nations' World Health Organization (WHO) and the Centers for Disease Control and Prevention (CDC) in the U.S. are good resources for research. See the Resources section, beginning on page 56, for contact information.

How is the red fire ant evolving?

The red fire ant has undergone at least one evolutionary change since arriving in North America: it has developed multiple-queen colonies. These polygyne colonies are unknown in the ant's native range. First reported in the U.S. in the 1980s, they are increasingly common, growing at a faster rate than single-queen colonies in the same range.

Polygyne colonies disperse in a new way, having evolved beyond the mating flight method (see page 17). The polygyne colony "buds," meaning a portion of its queens and workers split off to found a colony nearby.

The population density of polygyne colonies can exceed that of single-queen colonies by as much as six times. It's likely that the current spread of red fire ants is now a two-stage process. First, long distance colonists (new queens) establish a "beachhead" a kilometer or more beyond the existing mound. Then, these beachheads slowly expand, perhaps 30 meters per year, through budding.

With this evolutionary change yielding higher densities of red fire ants, whatever impacts the ants have on other species processes will be exacerbated. As an example, the native *S. xyloni* has apparently been eliminated from the parts of its range infiltrated by red fire ants.

FIGURE 1

The relationship between the geographic distribution and the mean skeletal size of P. domesticus, *the English house sparrow. Numbers 1–10 denote skeletal size from smallest to largest, and contour lines denote extended geographic range. Adapted from:* Evolution, *25:1-28, March 1971.*

FIGURE 2

Mean body weights of P. domesticus *plotted against isophanes (values for climatic conditions calculated from latitude, longitude, and altitude). Adapted from:* Science, *May 1964.*

Locations:

a. Houston, Texas
b. Los Angeles, California
c. Austin, Texas
d. Death Valley, California
e. Phoenix, Arizona
f. Baton Rouge, Louisiana
g. Sacramento, California
h. Oakland, California
i. Las Cruces, New Mexico
j. Lawrence, Kansas
k. Vancouver, British Columbia
l. Salt Lake City, Utah
m. Montreal, Quebec
n. Edmonton, Alberta

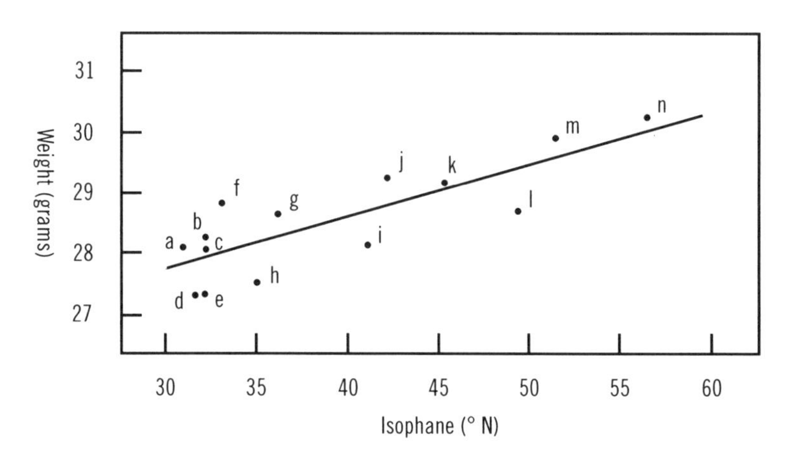

How does geographic location influence species size and shape?

Keeping warm in cold environments and cool in hot ones is important for survival. Birds and mammals that can regulate their temperature, such as by reducing body heat loss, have the best chance of thriving in frigid environments. This observation is the basis of Bergmann's Rule of Ecogeography: birds and mammals living in cooler environments will be larger than closely related species living in warmer environments.

Scientists have been debating the cause of this phenomenon since the 19th century, even though there is general agreement that a number of species, such as penguins, successfully illustrate this rule. The thermodynamic explanation for the Rule is that large animals waste less energy because they have a smaller surface-to-volume ratio than tiny animals.

A related observation is Allen's Rule of Ecogeography, which states that birds and mammals in cooler environments have shorter extremeties than their relatives in warmer ones. Arctic hares, for example, have shorter ears than their close relatives farther south. Thermodynamically, this prevents heat loss, and is one of several adaptations that prevents animals from freezing.

I. Pacific Northwest

II. Southwest

III. Midwest

IV. East Coast

V. Hawaii

FIGURE 3

Plumage variations among P. domesticus *individuals collected from five U.S. regions. Make a list of any variations in plumage that you observe. Be sure to keep accurate notes about which variations pertain to which bird, and record which region that bird is from. How might climatic and environmental conditions in a region account for a particular variation in plumage?*

Environmental Change and Controlling Impacts

OBJECTIVE

To quantify how an introduced species alters environments, and to explore control methods.

TIME MANAGEMENT

After initial set up, this activity requires periodic attention over the course of a week for Part A, with a final period allotted for class discussion. Part B may be completed in two or more periods, depending on how far you want students to carry out their control protocols.

BACKGROUND

Water hyacinth (*Eichhornia crassipes*) is an introduced species causing serious problems in many areas of the American Southeast, where extensive water hyacinth mats can sometimes cover entire lakes and rivers. In some areas, such as Florida, *E. crassipes* is under "maintenance control" thanks to long-term efforts of federal, state, and local environmental management agencies. Maintenance control means plant managers keep the plants at a relatively low level by using herbicides, machines, and biocontrol insects. Constant vigilance is required by aquatics management agencies to prevent *E. crassipes* populations from returning to infestation levels. If even as little as a year were to pass without continual maintenance by the couple of hundred aquatics management agencies in a particular state or region, *E. crassipes* populations would likely return to infestation levels. This would eventually require spending millions of dollars to return those populations to maintenance levels.

Elsewhere in the world, Uganda for example, this "world's number one" aquatic weed currently disrupts navigation, prevents fishing, threatens hydroelectric dams, and interferes with water supplies. Water hyacinths are native to South America, and their growth rate is among the highest of any known plant; *E. crassipes* populations can double in size in as little as 12 days. Besides blocking maritime traffic and preventing swimming and fishing, water hyacinth infestations also prevent sunlight and oxygen from penetrating down into the water column. A profusion of decaying hyacinth plant matter also reduces oxygen in the water. Thus, water hyacinth infestations dramatically alter the environments they inhabit, reducing fisheries, shading out submerged plants, crowding out emersed plants, and reducing biological diversity.

In this activity, students will first measure the impacts of water hyacinths on a controlled environment, then explore how various control techniques may be used to maintain water hyacinth populations at certain levels.

PART A
ENVIRONMENTAL CHANGE

Organisms interact with their environments and, in the process, cause changes to those environments. These changes can occur in a variety of ways, such as chemical changes in soils and water or changes in food webs and feeding relationships within an ecological community. Introduced species create new niches within a community, and can alter the competitive relationships among species.

In Part A of this activity, students investigate the different impacts of the water hyacinth on an environment by assessing water quality before and after the introduction of water hyacinths. Students also monitor the biological environment by surveying the microorganisms present in the aquarium both before and after the water hyacinths' introduction.

PROCEDURE

MATERIALS

Small aquaria or containers

Water hyacinths (*Eichhornia spp.*)

Thermometer

pH test kits

Dissolved oxygen test kits

Nitrogen test kits

Compound microscope

Glass pipette

Slides, coverslips

1. Set up two small aquaria, using a container, pond water, and water hyacinths, three days prior to this activity. Figure 1 (page 40) depicts an example of this set up. The aquarium container should be large enough to hold one liter of water. Large fish bowls, two-liter containers, empty plastic milk jugs, or other such containers may be used. Both containers used by a student group, however, should be of the same type. One container, without hyacinths, serves as the control in this experiment.

2. Have students assess the baseline water quality of the aquarium by measuring temperature, pH, dissolved oxygen, and nitrogen.

3. Have students assess the baseline community of microorganisms by sampling the protozoa present on the water hyacinth's undersides. To sample the protozoa present on the surface water, have students make quick estimates as follows. Using a glass pipette with a fine tip, take a water sample. Deposit a drop on a glass slide (about 0.01 ml). Add a coverslip, observe under low to medium power, and count the number of protozoa. Every student in the class should count a different sample taken from the same location; use the average of all the counts. Multiply the class average by 0.01 ml to get the total volume of water observed in milliliters. Divide the number of protozoa by the total volume of water to estimate the number of protozoa per milliliter.

4. Have each group discuss among themselves what they hypothesize will be the possible effects of introducing the aquatic plants at each of these baseline water quality parameters.

5. Have the groups form a hypothesis based on their discussion of the effects of the water hyacinths on each parameter, and design an experiment to test their group's hypothesis.

6. Over the course of a week, have each group collect data based on their experiment and record their findings in their journals. Have the groups share their findings with the class.

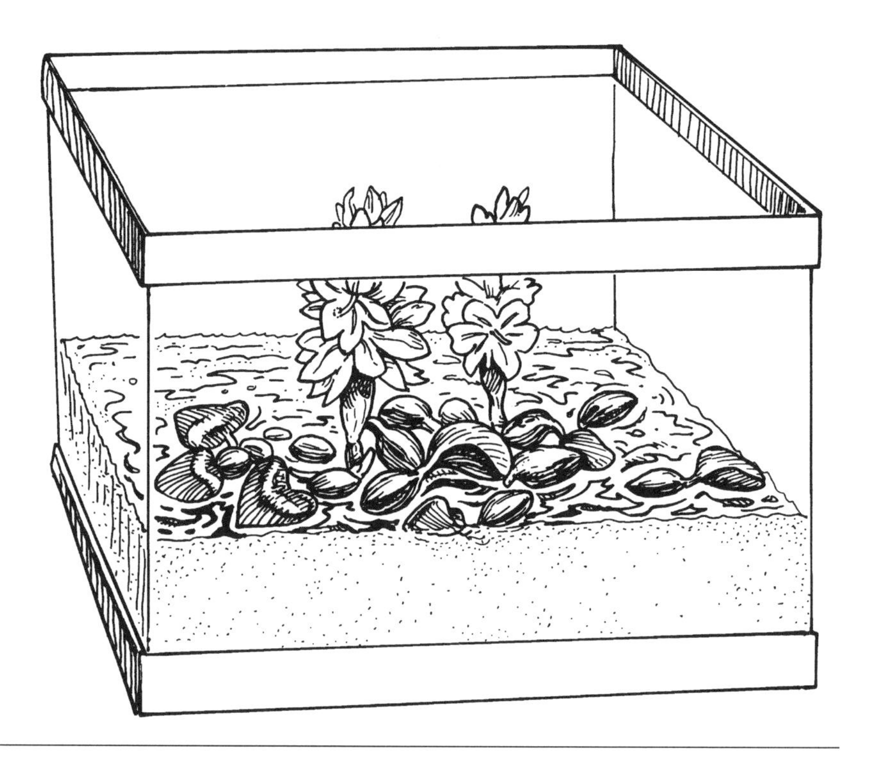

FIGURE 1

The aquarium container for this activity set up should be large enough to hold one liter of water. Other possibilities include large fish bowls and empty plastic milk containers. Both containers used by students should be of the same type and size.

Biocontrol insects and water hyacinths

Biological control, or biocontrol, is defined as controlling an introduced species with another species. The first biocontrol species released against water hyacinth populations in the U.S. was the mottled water hyacinth weevil (*Neochetina eichhornia*) in 1972. Both the adults and the larvae feed on various parts of the water hyacinth plant.

The second biocontrol insect released against water hyacinth populations in the U.S. was another *Neochetina* species, the chevroned water hyacinth weevil (*N. bruchi*), again in Florida but in 1974. Its life cycle is shorter than that of *N. eichhornia*, although its impact is similar. Another biocontrol insect used against water hyacinth populations has been the Argentine water hyacinth moth (*Sameodes albiguttalis*), whose life cycle is only 30 days. Only during its larval stage does it feed on the water hyacinth. It was released and is now established in Florida, Louisiana, and Mississippi, as well as in Australia, South Africa, and Sudan. It may retard water hyacinth growth in the early stages of mat development.

Not all biocontrol insects are from other geographic locations. The larval stage of *Bellura densa*, for example, is native to the southeastern U.S. where it is commonly called the pickerelweed borer. Efforts were made in the early 1980s to augment natural populations of its larvae in an effort to impact Louisiana water hyacinth populations. While these efforts were largely successful, increased moth larvae populations had little impact on Louisiana's water hyacinth communities.

Florida's water hyacinth populations are controlled through mechanical and chemical means, while the contribution of biocontrol methods is minimal. For further information on these and other biological control efforts, visit the University of Florida's Center for Aquatic Plants homepage at:

http://aquat1.ifas.ufl.edu/welcome.html

PART B
CONTROL

Action is often taken to prevent the invasion and spread of introduced species. But efforts at controlling the environmental impacts of introduced species can themselves cause changes and alterations in environments. Such changes must also be evaluated in relation to the overall environmental impact an introduced species may have. In Part B, students use water hyacinths to investigate the potential impacts control mechanisms can have on an environmental system.

PROCEDURE

MATERIALS

Aquaria with water hyacinths from Part A

Water testing and microscope equipment from Part A

Materials for one or more of the control protocols listed in the Procedure

1. Have students choose an experimental protocol from the list below for controlling the aquatic plants in an experimental aquarium.

   Protocol A
   Mechanical removal. Design a mechanical system for removing water hyacinths. Materials may include: straws, paper clips, string, and dowel rods.

   Protocol B
   Water level fluctuation. Develop a strategy for varying water level by either removing water (drought) or adding water (floods) to the aquarium system.

   Protocol C
   Physical control. Remove certain plant parts by chopping, ripping, or tearing them apart.

   Protocol D
   Chemical control. Add a chemical that will in some ways impede the growth of the aquatic plants (salt, baking soda, herbicides).

   Protocol E
   Biological control. Add an animal that will directly feed on the aquatic plants (snails).

2. For the protocol they have chosen, have students use their journals to write a hypothesis and design an experiment to test the effect of their protocol on the aquatic plants, protozoa community, and water quality of the aquarium system. Have students carry out their experiment, entering all data in their journals.

3. When experiments are concluded, have student groups design a poster explaining their control protocol and results. This provides an opportunity for the entire class to share and discuss its findings. Once the class has viewed and discussed all the posters, lead a general discussion to compare and contrast the various control protocols used by the groups. Which worked best? How could one or all of them have been made to work better? What might be some potential impacts of the control methods used if they were implemented among actual environments? Have students write two or three paragraphs in their journals summarizing their findings.

What control methods work against the red fire ant?

Over $200 million has been spent so far in the United States to control the red fire ant. The early centerpiece of this attempt was an eradication campaign by the U.S. Department of Agriculture that has been called "the Vietnam of entomology" by noted entomologist E.O. Wilson.

In 1957, Congress authorized $2.4 million to eradicate the species, but the heptachlor pesticide that was used killed cattle and wildlife, and effects on human health were not carefully monitored. Researchers next developed baits of Mirex, another pesticide, but traces of Mirex soon showed up in many non-target species, including humans, and the registration of Mirex was canceled by the U.S. Environmental Protection Agency in 1977, effectively terminating the campaign. During the eradication campaign, the range and density of the red fire ant increased greatly, often at the expense of native ants. Because the red fire ant is a weedy species (so called because it reproduces and disperses so well), any area cleared of all species of ants by pesticides was more rapidly recolonized by the red fire ant than by native ants.

Currently, there is no effective way to control fire ants over large areas, although various baits and insecticides are still used. A homeowner can easily kill individual colonies by pouring large amounts of boiling water on the mounds, but this approach is hardly practical for a farm or other large area. Many land owners are interested in introducing natural enemies of the red fire ant from its native range in Brazil, in the hopes that one or more of these will control its population.

This biocontrol approach has been successful for a number of exotic agricultural pests, although it has more often failed to control the target species and has sometimes led to the control agent attacking non-target native species, even some of conservation concern. So, before additional Brazilian species are introduced, researchers need to learn possible targets of the control agent.

Bacteria, fungi, protozoa, parasitic nematodes, and even an ant whose queens are parasitic in red fire ant colonies have been tested as potential biological controls. The major current project culminated in the release, in 1997, of Brazilian phorid flies in Florida. These flies lay a torpedo-shaped egg inside the ant's body. When the larva hatches, it eventually crawls to the ant's head and causes it to fall off. The larva then uses the ant head as a shelter and grows into an adult fly.

Whether this fly will achieve significant control of the red fire ant in its new range, or simply be another in a line of failed, well-publicized attempts to master this invader, will become clear in the next few years.

SUGGESTIONS FOR FURTHER STUDY

Students have learned in this activity that biological control is one of several mechanisms used to offset the harmful impacts of introduced species. To prepare students for the culminating activity, Balancing Human Systems with Natural Processes, have them research other methods used to control impacts of introduced species. Both ecological restoration and outright eradication, for example, have their own sets of pluses and minuses when used as control measures. Encourage students to develop an understanding of how cost/benefit and risk analyses are used to establish and implement control measures.

Students need to learn how to ask specific, focused questions when analyzing environmental and other issues. In this way, the answers they derive will be focused, and they will be based on scientific data and other forms of credible, useful evidence. For example, if a particular introduced species—such as the brown tree snake, *Boiga irregularis*, studied in the culminating activity—is perceived to pose a hazard to human systems and ecological processes, what exactly is meant by the term "hazard"? Begin this discussion by telling students that a hazard is something that has the potential to cause harm or undesirable consequences to humans or what they value. Alternatively, have students look up the term in a dictionary or similar resource. Then have students address the questions outlined below. Encourage them to use a species they have been studying as they do so. In this way, students will both learn how to pose focused questions and develop an understanding of what a "hazard" is.

Who or what is exposed to the hazard? Is it humans, non-humans, ecological processes? Is it something humans value, such as natural beauty, valuable property, or a built system?

Where is the hazard experienced? Is it confined to one area, or is it regional or global? If confined, does it have the potential to spread?

What is causing the hazard? With an introduced species, for example, is its population growing faster than the surrounding ecosystem's ability to absorb it? Is it crowding out or eradicating other species? Are humans being hurt or harmed in some way? If so, how? Is there a social or ethical component, such as one group of people is harmed while another benefits? Are social, community, or political relationships affected, either negatively or positively?

What is occurring now, and what will occur in the future? Can scientists accurately measure current circumstances and use the data they compile to form predictive hypotheses? If not, what can be done to improve data collection techniques? What might happen if nothing is done? What are some best- and worst-case scenarios?

Have students record the results of this discussion in their journals for reference during the culminating activity.

Balancing Human Systems with Natural Processes

OBJECTIVE

To role-play a National Conference on Introduced Species that will determine an appropriate response to the invasion of the brown tree snake in Oahu, Hawaii.

BACKGROUND

Students began their exploration of introduced species by examining differences among native, introduced, and invasive species (Activity 1). In lab investigations and studies of particular invasive species, they have acquired tools and skills for understanding dispersal (Activity 2), competition (Activity 3), hybridization (Activity 4), evolution (Activity 5), and chemical change and biological control (Activity 6). In carrying out their investigations, students have explored the significant and complex interactions among the different branches of science—especially biology and ecology—that must be factored into any "real-world" investigation of the impacts of introduced species. Students have also learned that a clear understanding of such impacts requires a broad range of scientific information, knowledge including an organism's physiology, its reproductive capabilities and population growth dynamics, its behavior and adaptability, and its biochemical interactions with new surroundings.

This culminating activity enables students to design a response to the threat of introduced species in Hawaii. Students role play to determine how to respond to the impact of an invasive species, the brown tree snake (*Boiga irregularis*), recently introduced to the island of Oahu, Hawaii. In a simulated National Conference on Introduced Species, students will use this real-world situation to demonstrate and extend their scientific knowledge.

Students will begin by familiarizing themselves with the materials contained in the National Conference Information Packet. They will then discuss these materials as a class, guided by the knowledge they've acquired during these activities and their other investigations. After assuming the roles provided, students will conduct a conference to decide on an appropriate response to the "invasion" of the brown tree snake.

PART A
IDENTIFYING THE PROBLEM

PROCEDURE

MATERIALS

Photocopies of National Conference on Introduced Species Information Packet (pages 50-55)

1. Distribute photocopies of the National Conference on Introduced Species Information Packet (pages 50-55) to student groups. A black line master is provided; you may make additions appropriate to your class. For example, add appropriate dates to personalize the conference, or include discussion of or quotes from any guests or experts with whom students have interacted during previous or other activities.

2. Explain that the conference's goal is to determine policy that will address the introduction of an alien species to Hawaii. With each student playing a different role, the class will debate public policy related to the impact of the brown tree snake on natural processes and human systems. You may wish to distribute copies of the scoring rubric (page 49) so that students will know how they will be assessed.

3. Have the class read the *Washington Post* article and the two letters in the National Conference Information Packet. This may be done as homework or in class, depending on the time you have allotted.

4. After students have read the packet, lead a class discussion to build a base of understanding. Summarize the articles, identify the problems they describe, and brainstorm possible controversies and solutions. The following questions and answers may be used to focus the discussion:

Q: Why was the appearance of a 1.5-meter snake at Oahu's Hickam Air Force Base of extreme concern to federal and state wildlife officials?
A: Hawaii has no typical snakes. One variety, the Brahminy blind snake (*Ramphotyphlops braminus*) is native to Hawaii, but its length seldom exceeds 150 mm and it poses no known hazards.

Q: Why is Hawaii particularly vulnerable to invasive species such as the brown tree snake?
A: Hawaii is isolated. Animals native to Hawaii have evolved with few natural predators or diseases, and therefore have few natural defenses. There are no natural predators of the brown tree snake in Hawaii.

Q: In Hawaii, what efforts can control snake importing by pet owners?
A: Large fines and jail time for illegal snake importation. Amnesty program for owners who turn their snakes over to authorities.

Q: What measures can control the invasion of the brown tree snake?
A: Concrete barriers erected around the airport tarmac where cargo is offloaded, dog detection, snake traps, fumigation of cargo containers, training of snake experts, biological controls such as viruses and parasites, chemical fertility inhibitors, and public education.

Q: Why have the brown tree snakes discovered in Hawaii focused attention on the brown tree snakes in Guam?
A: The brown tree snake is a major problem in Guam, causing power outages and extinction of many native species. People fear a similar problem could develop on the Hawaiian Islands.

PART B
PREPARING FOR THE NATIONAL CONFERENCE
PROCEDURE

1. Have students research their roles. Research sources include on-line materials and students' own journals. Groups of students with similar roles might work together. When they finish researching, ask students to frame a position statement according to the role they are playing. Guide students to address the following in their statements:
   - What are the biological impacts of their solutions?
   - What are the ethical aspects of these solutions?
   - What are the ecological impacts of these solutions?

2. As a summary method, each group should make a poster with a one-sentence description of the group's goals and research needs. Leave space for each member to add a specific plan and research direction that further focuses on their needs. Students should prepare these in advance, and mount them on the poster during their presentations.

3. When the conference is convened, each student will give a one-minute talk, discussing their policy statement and the research needed to ensure their proposed policy is effective. You might also encourage students to come to the National Conference in costume. Have each student make a name tag that gives her/his role.

4. A week or two in advance, you may wish to invite other teachers, the principal, school board members, local biologists, or a reporter from a local paper to attend the Conference. They can attend either as participants (speaking from their own point of view) or as impartial observers who vote on the policy to adopt.

PART C
THE NATIONAL CONFERENCE

PROCEDURE

1. Before class, you might set up a video camera to tape the conference. Students usually enjoy watching themselves in action, and it provides you with a record if you need to review something a particular student said. If possible, seat students in a circle so they can see each other and the posters. Remind students to stay in character.

2. You may wish to establish the criteria on which you will reach a decision before the conference begins. The following criteria will begin this process:
   - Which plans address the most important needs for Hawaii's human population?
   - Which plans address the needs of the greatest number of threatened species?
   - Which plans have the greatest probability of success?
   - Of all the ideas presented, which solution best balances natural processes with human activity?

3. While each student should present her/his one-minute talk individually, it may be useful to have the different groups present their testimony as a whole. Group members will probably discuss policy and research needs from a similar viewpoint; their combined presentation should give a realistic sense of the group's goals and suggestions.

4. After the presentations have been made, the class will stay in their roles to discuss what to do about the problem of the brown tree snake. (Depending on time constraints, this discussion might need to take place the day after the presentations are given.) The amount of time spent on the discussion is flexible; it can be as long or as brief as necessary. You may either choose to moderate the discussion yourself, or appoint a participant to moderate.

5. Have the class stay in character and vote on the plan or plans that the conference suggests should become standard policy. Then ask the class to step out of their roles and vote again. Do the results differ? Ask any students who changed their votes to explain why they did so.

6. At the end of the debate, have participants turn in copies of their opening remarks and other pertinent materials.

7. You will probably find it helpful to use the scoring rubric on the opposite page to help in evaluating student presentations. If you decide to have students critique their own efforts, as in a peer review, you might want to photocopy the scoring rubric for inclusion in their National Conference Information Packets.

PROCEDURAL NOTES

Students should be informed in advance that they will be asked to give a one-minute presentation on the solutions they think provide the best balance between natural processes and human activity in the case of the brown tree snake invasion of the Hawaiian Islands. This information will guide them in their research, help them develop appropriate presentation formats, and inform their debate. You may find it useful to have students provide you with a brief outline of the content of their intended presentations before they make them. This will not only help you orchestrate this portion of the activity but, depending on the amount of time you've allotted, may encourage students to revise their original presentations according to your direction and advice. It will also enable you to identify and help any students who might need assistance before they address the class.

ASSESSMENT NOTE

Many opportunities for assessment and evaluation present themselves during the course of this activity. One of the most significant, however, is how well students listen to one another and work together. In the "real world" of environmental policy debate, conflict resolution often involves compromise.

Students should be encouraged to keep notes while speakers make their presentations, and to write down any questions and concerns they may have for use in the debate that follows. They may do this in their student journals. It may prove best to have students save all their questions for the debate portion of the activity, but you can also hold a more informal question-and-answer session following either each speaker's presentation or all the presentations. This may help students address any questions and concerns as they arise. Regardless, students should be reminded to be polite and considerate to one another at all times, and to bear in mind that they are participating in a democratic decision-making process.

After their presentations are completed, conference participants will then take part in a moderated debate to address the items listed below. You may want to write these items on the chalkboard. As the debate unfolds, have students add to their existing notes in their journals.

1. What are the possible ways to address the situation?
2. What are the biological impacts of these solutions?
3. What are the ethical aspects of these solutions?
4. What are the ecological impacts of these solutions?
5. Of all the ideas presented, which solution best balances human activity with natural processes?

SCORING RUBRIC

0–4 points will be given for:
- speech prepared
- questions and responses to questions
- dramatic nature of role play
- statement of goals
- research plan to support the goal

Level 0
Little evidence of preparation. Use of points made from previous study are rare or nonexistent. Frequently distracted or disruptive. Does not come in costume. Does not understand the role that he/she is playing. No evidence of outside research.

Level 1
Speech prepared but missing major points or speech not characteristic of this person's role. Use of one or two of the major points presented in previous study. Attentive, but no evidence of active participation in the follow up questions. Does not come in costume. Frequently steps out of character. No evidence of outside research.

Level 2
Speech well prepared but lacking in logic. Use of some of the major concepts presented in previous study. Listens to the questions of others and is attentive to the conference proceedings. Does not initiate questions and has difficulty answering questions that are addressed to him/her. Occasionally steps out of character. No evidence of outside research.

Level 3
Speech well prepared, logically presented, use of some of the major concepts presented in this previous study. Usually on task, listens well to questions of others. Answers questions that are addressed to him/her directly. Does not initiate many questions. Evidence of some outside research. Stays in character.

Level 4
Speech well prepared, logically presented, use of a large number of major concepts presented in previous study. Attentive to questions asked of all participants, answers to questions well thought out, gives evidence of listening to all peoples' speeches, questions, and answers. Evidence of much outside research. Comes in costume and stays in character.

A LETTER FROM THE NATIONAL CONFERENCE ORGANIZERS

Aloha, Colleague:

We're pleased you will be joining us for the National Conference on Introduced Species, which will focus on the brown tree snake's invasion of Hawaii. Before we begin, we'd like to remind everybody to keep an open mind; there are no perfect answers. There will be drawbacks to any approach the conference decides to take, so our purpose is to hear the evidence from all perspectives, to hear and discuss participants' ideas for addressing this situation, and to recommend a plan of action. Conference participants come from very different backgrounds, so we recommend you read through this packet—to learn the basic aspects of all sides of the situation—before researching the subject from your own area of expertise.

Recent reports from Hawaii indicate that the brown tree snake (*Boiga irregularis*) has been sighted in the cargo area of Honolulu's Hickam Air Force Base. On the surface, this would not seem to be a particularly unusual occurrence, except that Hawaii does not have any typical snakes. Moreover, the cargo plane from which the snake emerged came from Guam, an area in which the brown tree snake has caused untold ecological and economic damage, including the extinction of both native and non-native birds.

Citizens and wildlife biologists are concerned that a similar invasion by this aggressive, poisonous, predatory reptile may be occurring in Hawaii. "If we can eliminate the snakes before its population explodes, perhaps we can head off its spread," they claim. "In this way, we may be able to prevent the snake from getting to the mainland." Several groups have asked the U.S. government to take steps to eradicate this invasive species before its population grows to a size that will damage the environment.

The U.S. Fish and Wildlife Service was prepared to bring to Hawaii some snake population control measures that were successful on Guam. However, concerns have been raised about the ethical and ecological consequences of eradicating one of nature's species. Said one distinguished U.S. senator, "Killing nature's creatures is an abomination. We humans have no right to interfere. As of now, I'm inclined to ban the killing of snakes altogether."

Additional concerns were voiced by Hawaiian farmers, who are concerned that funds spent fighting the brown tree snake will decrease the money available to fight fruit flies. Fruit flies in Hawaii destroy an estimated $300 million in crops every year. Also, lobbyists for several national home insurance companies are requesting the subcommittee consider the costs of ignoring yet another introduced species, the Formosan ground termite. In response to these objections, the U.S. Fish and Wildlife Service has temporarily suspended its snake eradication program.

To encourage discussion of these important perspectives, conference organizers decided that Hawaii should be the focus of a National Conference on Introduced Species. Participants will include leaders in the fields of ecology, biology, evolution, and chemistry. Participants will also include governement leaders, such as the governor of Hawaii, and citizens who would be affected by any decisions.

We invite you to use every possible resource as you research the brown tree snake situation. The National Conference will convene on __________, at precisely ____am/pm. Please prepare a one-minute paper to discuss the research supporting your decision. After a break, the Conference will reconvene on ___________, at precisely ____am/pm, to debate the issues raised by all participants.

Mahalo,
Organizers of the National Conference on Introduced Species

AN OPEN LETTER FROM THE TERRITORY OF GUAM

To the National Conference participants:

This letter serves as a warning to any community into which the brown tree snake has been introduced. As a researcher in Guam, I have seen the catastrophic impacts this species has had on our island. The absence of natural population controls and the plentiful food supplies on Guam have turned the brown tree snake (*Boiga irregularis*) into an exceptional pest that causes major ecological and economic problems.

B. irregularis arrived on Guam in ship cargo from the Admiralty Islands, about 1750 kilometers to the south. The first sightings on Guam were inland from the seaport in the early 1950s, and they became conspicuous throughout central Guam by the 1960s. By 1968, they had probably dispersed throughout the island. In its native habitat (Solomon Islands, New Guinea, and northern Australia), *B. irregularis* has quite different growth, survivorship, and density patterns, and therefore does not cause the problems seen here.

B. irregularis has had and continues to have serious economic impacts here. It preys on farm animals, poultry, and pets. It has caused the loss of pollinators for Guam's forest trees by consuming many insects, and has virtually wiped out Guam's native forest birds. Twelve species of birds, some found nowhere else, have disappeared from the island, and several others are precariously close to extinction. Some introduced birds better adapted to resist snake predation persist on Guam only in urban areas and other developed sites where snake density is limited by human activity and inappropriate habitat conditions.

Snakes crawling on electrical lines frequently cause power outages and damage electrical lines maintained by Guam Power Authority and Naval Public Works. Since 1978, more than 1200 power outages have been caused by snakes, and the resulting damage to electrical equipment is a significant economic burden to nearly all civilian and military activities on Guam. The power interruptions cause a multitude of problems, ranging from food spoilage to computer failures.

Total economic impacts resulting from endangered species constraints on island development are incalculable. Ecological impacts are primarily the interaction of *B. irregularis* with its prey and other species.

Over 8,000 snakes per square kilometer may live in some forested areas of Guam. Many snakes are killed by automobile traffic and island residents. Others are electrocuted while climbing on electrical lines. But, despite this mortality, *B. irregularis* abundance levels remain high in nearly all forested and urban habitats throughout Guam. Pigs and monitor lizards do eat snakes, including *B. irregularis*, but they are ineffective in lowering the brown tree snake population in most areas of the island.

A variety of groups are researching methods for containing this species. As a resident of Guam, I urge you to consider their research, as well as this species' impact on our island, before deciding how you will address this issue.

Sincerely,
Madeleine Gutierrez

Source: This letter was developed using resources from: USGS-Biological Resources Division, Patuxent Wildlife Research Center, National Museum of Natural History, Washington, D.C. 20560-0111; and Tree Snake Online, a joint project of the USGS Biological Resources Division & the Pacific Basin Development Council.

Serpentless Hawaii Fears Snake Invasion

BY WILLIAM CLAIBORNE
Washington Post Staff Writer

HONOLULU— Shortly after a huge transport plane unloaded its cargo at Hickam Air Force Base one day earlier this month, Airman John Herist happened to spot a brownish, three-foot-long snake slither into a nearby canal and disappear.

An unremarkable event by almost any measure, except that Hawaii does not have snakes and the cargo plane was from Guam, a combination of circumstances that had state and federal wildlife officials scurrying to set traps and turn loose snake-sniffing Jack Russell terriers in a frantic round-the-clock hunt for the elusive reptile, which still has not been found.

Brown tree snakes are an aggressive, venomous predator that grows to lengths of eight feet and has spread throughout Guam like a plague since arriving aboard U.S. military cargo ships from the Solomon Islands shortly after World War II. They now number 12,000 per square mile in some forested areas of the Pacific island and are eating into extinction its native bird species and most of the non-native birds as well.

Now officials here are worried that the brown tree snake, hiding in aircraft cargo holds and wheel wells, may be invading Hawaii, threatening its wildlife habitat and tourism-dependent economy. More than a third of all the threatened and endangered birds in the United States are found in Hawaii.

A nocturnal reptile with a large head and bulging eyes, the brown tree snake prefers birds over other prey, but it has been known to eat small pets such as cats and dogs and has even been found curled around babies sleeping in their cribs. It also crawls along electrical lines and causes an average of one power outage every four days on Guam.

Hawaiian wildlife officials say that while there have been only seven confirmed cases of brown tree snakes being killed or found dead on Hawaii's Oahu island since 1981, the Hickam Air Base incident was the sixth snake sighting in two months. They also warn that even one pregnant female slipping through could begin a colonization far more costly than Guam's.

"It's an enormous threat to Hawaii, and while we always look for the 'silver bullet' to kill these things off, we haven't found one yet," said Robert Smith, Pacific Islands manager for the U.S. Fish and Wildlife Service. "We've got to apply resources to this effort that match the cost of this threat."

Because of its isolation, Hawaii is particularly vulnerable to invasive species like the brown tree snake, wildlife experts say. Animals here evolved with few diseases and natural predators, and therefor have few natural defenses. There are no effective predators with which the brown tree snake would have to contend while it multiplied.

But the threat is not only to Hawaii, according to U.S. Agriculture Department officials. One brown tree snake was found in a cargo in Texas, and experts predict that the reptile could easily thrive in Southern California, Florida, and other warm climate states.

Thomas H. Fritts, a biologist with the U.S. Geological Survey in Washington who is widely regarded as the leading authority on brown tree snakes, said he was attempting to confirm sightings in Spain, Singapore, Okinawa, and Darwin, Australia. An incipient colonization is already occurring in Saipan, he said.

The sighting at a military airfield in Darwin was ironic because Australia is where the reptile is believed to have originally evolved before island-hopping through the Pacific to Guam. "So we seem to have moved it from Guam back to Australia," Fritts said.

Because Hawaii ostensibly has no snakes—other than two reptiles on display in the public zoo here and those illegally imported by residents who like to have them as pets—state and federal officials take their snake control efforts seriously, even though the state's congressional delegation is often the butt of jokes when it lobbies for appropriations for alien snake control programs.

Anyone caught with a snake faces as much as a

(Reprinted with permission. *The Washington Post*, August 23, 1997. "Trouble in Paradise? Serpentless Hawaii Fears Snake Invasion," by William Claiborne. p. A1.)

year in jail and a maximum fine of $25,000. An amnesty program allows snake owners to turn the reptiles in without prosecution.

In addition, a Coordinating Group of Alien Pest Species, comprised of 14 government agencies and private groups, last year drafted a 10-point "Silent Invasion" action plan to improve alien pest prevention and control programs. It includes a brown tree snake control plan that will be boosted by nearly $1.8 million in federal appropriations this year for combating the reptile on Guam, researching new control methods and inspecting aircraft arriving in Hawaii.

A number of measures have been taken or proposed to intercept snakes that arrive from Guam in military aircraft or other conveyances. These include a newly designed concrete barrier with a curved lip that could be erected around an airport tarmac where cargo is off-loaded.

Other measures include dog detection, which is used extensively in Guam; development of new kinds of snake traps; fumigation of cargo containers, and the training of snake searchers who are marshaled when a brown tree snake is spotted loose. Experts also called for more research on chemical fertility inhibitors and the use of toxicants and viruses and parasites known to be exclusively effective against the brown tree snake.

But experts said the problem with biological controls like viruses is that they have not worked well on vertebrate species and unless an entire species population is wiped out, the survivors may develop immunities and recolonize. Also, they said that other nonthreatening species of wildlife might be affected by pathogens.

High priority is also being given to public education. Even through most Hawaiians have never seen a snake, Smith said that "people here hate snakes and what might seem to be preaching to me choir really gets the congregation rattled."

"We've got to apply resources to this effort that match the cost of this threat." —Robert Smith

Earl Campbell, a wildlife ecologist with the USDA's National Wildlife Research Center, said each of these control measures may be effective to a point, but he stressed that only "a gamut of techniques" used in concert will stand a chance of preventing the brown tree snake from colonizing here as it did in Guam.

"Time is of the essence. With the unique ecosystem we have in Hawaii, we can't afford to delay. It could be disastrous for us," Campbell said.

David Worthington, a U.S. Fish and Wildlife biologist here, said that Guam has to be the focus of much of the brown tree snake control effort because "if we don't find a way to reduce the population there, everything else is just stopgap."

Fritts said that while the explosive population growth of brown tree snakes in Guam is reversible, "I don't think we will ever eradicate it entirely there. Besides, the problem is not just in Guam. You don't stop yellow fever in the United States by just wiping it out in Brazil. We won't get anywhere if we don't appreciate how complex these infestations are."

While the brown tree snake is regarded by wildlife officials as the most imminent alien species threat to Hawaii, it is not the only one, or even the worst. And efforts to control it must compete with programs for controlling other invasive pests.

Each year, an average of 20 new insects arrive here, half of them known pests. Fruit flies cost Hawaii farmers an estimated $300 million a year, and the Formosan ground termite costs nearly $150 million a year in treatment and repair of damage to houses.

In 1995, biting sand flies, which can inflict up to 10,000 bites per human in a day, were discovered—and subsequently destroyed—aboard three canoes that made a commemorative voyage from the Marquesas to Hawaii. Three years earlier, quarantine officials confiscated 39 flesh-eating piranhas shipped by a mainland mail order house to aquarium enthusiasts here, and in 1991 red fire ants almost maid there way to Hawaii when agriculture inspectors intercepted a mail parcel from Florida.

But the brown tree snake is beginning to attract the attention it deserves, said Alan Holt, deputy director of the Nature Conservancy here. When President Clinton visited here last October, Holt said, wildlife officials told him that the invasive reptile was one of the three top economic issues in the state and that the president promised, "I'm going to help you on this." The appropriation for the control program followed almost immediately.

Nonetheless, Hold said the brown tree snake still is a "head scratcher" for many in Congress "who just don't understand what all this fuss over snakes is about."

"But the brown tree snake has taken us to a new level. No one has ever mounted a population control measure to wipe out an entire species of snakes before," he said. "We have our work cut out for us."

LIST OF NATIONAL CONFERENCE PARTICIPANTS

Listed below are the four main groups of concerned citizens participating in the conference, along with brief descriptions of the individual participants who have identified themselves as members of each group (according the the pre-conference survey). Each group will have very different opinions about the brown tree snake invasion of Hawaii situation. Because these are general groupings, individual participants within a particular group may even have different opinions.

GROUP 1

CITIZENS OF THE ISLAND OF OAHU

Participants include: a worker at Oahu Central Power Company; a farmer with an infant; a grocery store owner; a pet store owner who is also a snake enthusiast; a member of a group that ethically objects to killing animals; and an insurance agent whose company covers costs for repairing houses damaged by Formosan termites and other pests.

GROUP 2

CITIZENS OF GUAM

Participants include: a representative from the Guam Power Authority; a resident of a forested area of Guam; and a brown tree snake researcher currently doing field studies in Guam.

GROUP 3

LOCAL AND FEDERAL OFFICIALS

Participants include: Hawaii's minister of tourism; the President of the Honolulu Chamber of Commerce; the Commanding Officer of Hickam Air Force Base; and the Governor of Hawaii.

GROUP 4

GOVERNMENT AGENCY REPRESENTATIVES AND SCIENTISTS

Participants include: two members of the Coordinating Group of Alien Pest Species (a group of government agencies and private organizations); a biologist who has been studying brown tree snakes in Guam; an expert on biological control; a biologist who specializes in the Japanese beetle; and an ecologist who studies species dispersal.

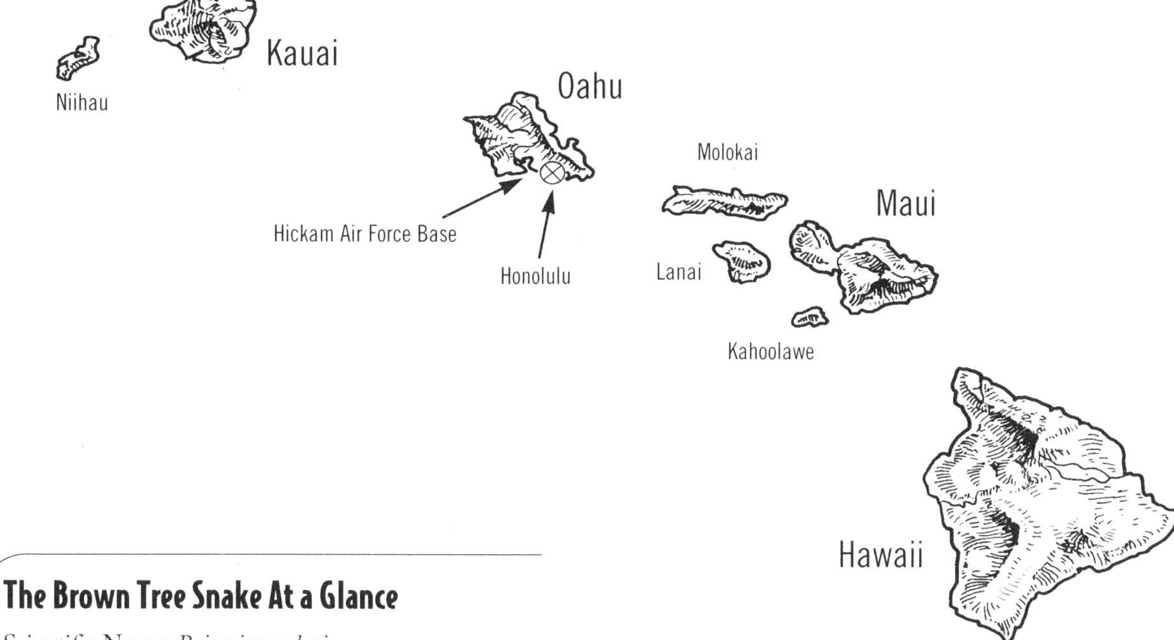

The Brown Tree Snake At a Glance

Scientific Name: *Boiga irregularis*

Origin: Northern Australia, Indonesia, Papua New Guinea, and Solomon Islands. First seen in Guam in the 1950s. Probably introduced through military cargo transport.

Diet: Wide variety of birds, small mammals, and lizards.

Size: Adults can reach 2.5 meters long and 2.3 kilograms.

Reproduction: Deposits up to 12 eggs as often as twice a year.

Threat to humans: Mildly venomous. Infants and small children could be harmed through suffocation or venom.

Source: U.S. Geological Survey, Biological Resources Division.

Resources List

NSTA Publications

NSTA offers many publications for teaching about global change and the impact of human activity on global systems. For sales: (800) 722-NSTA (U.S. & Canada). For general information: (703) 243-7100. Publication sales and information may also be obtained via NSTA's home page: http://www.nsta.org.

Anderson, O. and Druger, M. (eds.). 1996. *Explore the World Using Protozoa.* Grades 9-12, 224 pp., PB# 137X.

National Science Teachers Association. 1998. *A Science Educator's Guide to Assessment.* Grades 9-12, 220 pp., PB# 145X.

National Science Teachers Association. 1997. *Decisions—Based On Science.* Grades 9-12, 144 pp., PB# 141X.

National Science Teachers Association. 1996. *Global Environmental Change: Biodiversity.* Grades 9-12, 64 pp., PB# 138X01.

National Science Teachers Association. 1997. *Global Environmental Change: Carrying Capacity.* Grades 9-12, 64 pp., PB# 138X03.

National Science Teachers Association. 1997. *Global Environmental Change: Deforestation.* Grades 9-12, 64 pp., PB# 138X02.

Rhoton, J. and Bowers, P. (eds.). 1996. *Issues in Science Education.* General, 228 pp., PB# 127X.

Scripps Institution of Oceanography. 1996. *Forecasting the Future: Exploring Evidence for Global Climate Change.* Grades 6-10, 160 pp., PB# 118X.

The Science Teacher

The Science Teacher is NSTA's award winning professional journal for teachers in grades 7–12. This monthly publication features articles written by educators on a wide range of scientific topics, innovative teaching ideas and experiments, and current research news. *The Science Teacher* is one of many benefits of NSTA membership. For membership information, call (800) 830-3232 (U.S. & Canada) or (703) 243-7100. The following articles from *The Science Teacher* relate to the *Global Environmental Change* series:

Hassard, J. and Weisburg, J. "The Global Thinking Project." April 1992, pp. 42–47.

Singletary, T. and Jordan, J. "Exploring the GLOBE." March 1996, pp. 36–39.

NSTA-Marketed Publications

NSTA's Science Store markets the following books produced by other publishers on related subjects. This list is subject to change; contact NSTA for updated information.

Barton, J. and Collins, A. (eds.). 1997. *Portfolio Assessment: A Handbook for Educators.* Grades K-12, 120 pp., OP# 486X.

International Council of Scientific Unions. 1994. *Global Change.* Grades 10–College, 320 pp., OP# 463X.

Ohio Sea Grant Education Program. 1995. *Great Lakes Instructional Materials for the Changing Earth System: Integrative Activities on Global Change.* Grades 5–12, 203 pp., MS# 240X.

Books & Articles

Atkinson, I. 1989. "Introduced Animals and Extinction," in *Conservation for the 21st Century*, Western, D. and Pearl, M. (eds.). New York: Oxford University Press.

Claiborne, W. 1997. "Trouble in Paradise? Serpentless Hawaii Fears Snake Invasion," in *The Washington Post*, August 23. p. A1.

Collard, S. 1998. *Alien Invaders: The Continuing Threat of Exotic Species.* Chicago: Franklin Watt.

Crosby, A. 1986. *Ecological Imperialism: The Biological Expansion of Europe, 900–1900.* Cambridge (UK): Cambridge University Press.

Fritts, T. and Rodda, G. 1996. "Trouble in Paradise: the Brown Tree Snake in the Western Pacific," in *Aquatic Nuisance Species Digest*, vol. 1, no. 3. pp. 26-27.

Mooney, H., Drake, J., *et. al.* (eds.). 1989. *Biological Invasions: A Global Perspective.* Chichester (UK): Wiley & Sons.

Mooney, H. and Drake, J. (eds.). 1986. *Ecology of Biological Invasions of North America and Hawaii.* New York: Springer–Verlag.

Office of Technology Assessment, U.S. Congress. 1995. *Harmful Non-Indigenous Species in the United States.* Washington, DC: U.S. Government Printing Office.

Long, J. 1981. *Introduced Birds of the World: The Worldwide History, Distribution, and Influence of Birds Introduced to New Environments.* New York: Universe Books.

Mlot, C. 1997. "Biological Pest Control Harms Natives," in *Science News*, vol. 152. p. 100.

Rhymer, J. and Simberloff, D. 1996. "Extinction by Hybridization and Introgression," in *Annual Review of Ecological Systems*, vol. 27. pp. 83-109.

Rodda, G., Fritts, T., and Conry, P. 1992. "Origin and Population Growth of the Brown Tree Snake on Guam," in *Pacific Science*, vol. 46. pp. 46-57.

Roughgarden, J., Heckel, D., and Fuentes, E. 1983. "Coevolutionary Theory and the Biogeography and Community Structure of *Anolis*," in *Lizard Ecology: Studies of A Model Organism*, Huey, R., Pianka, E., and Schoener, T. (eds.). Cambridge: Harvard University Press. pp. 371-410.

Ruibal, R. 1965. "Evolution and Behavior in the West Indian *Anoles*," in *Lizard Ecology: A Symposium.* Milstead, W. (ed.). Columbia: University of Missouri Press. pp. 116-140.

Simberloff, D. 1996. "Impacts of Introduced Species in the United States," in *Consequences: The Nature and Implications of Environmental Change*, vol. 2, no. 2. pp. 13–22.

Simberloff, D. 1995. "Introduced Species," in *Encyclopedia of Environmental Biology, vol. 2.* San Diego: Academic Press.

Simberloff, D., Schmitz, D., and Brown, T. 1997. *Strangers in Paradise: Impact and Management of Nonindigenous Species in Florida.* Washington, DC: Island Press.

Williams, E. 1983. "Ecomorphs, Faunas, Island Size, and Diverse End Points in Island Radiations of *Anolis*," in *Lizard Ecology: Studies of A Model Organism*, Huey, R., Pianka, E., and Schoener, T. (eds.). Cambridge: Harvard University Press. pp. 326-370.

Williamson, M. 1996. *Biological Invasions.* London (UK): Chapman & Hall.

Wilson, E. 1988. *Biodiversity.* Washington, DC: National Academy Press.

U.S. Government and International Agencies

Environmental Information Service
National Oceanic and Atmospheric Administration (NOAA)
1315 East-West Highway, Room 15400
Silver Spring, MD 20910
(301) 713-0575
http://www.esdim.noaa.gov

Natural Resources Conservation Service
1400 Independence Avenue, SW
Washington, DC 20250
(202) 720-2791
http://www.ncg.nrcs.usda.gov

U.N. Development Programme
1755 K Street, NW
Suite 420
Washington, DC 20006
(202) 331-9130
http:/www.undp.org

U.N. Environment Programme
Office of the Executive Director
PO Box 30552
Nairobi, Kenya
http://www.unep.no

U.S. Department of Agriculture
1400 Independence Avenue, SW
Washington, DC 20250
(202) 720-2791
http://www.usda.gov

U.S. Department of the Interior
1849 C Street, NW
Washington, DC 20240
(202) 208-3100
http://www.doi.gov

U.S. Environmental Protection Agency
401 M Street, SW
Washington, DC 20460
(202) 260-2090
http://www.epa.gov

U.S. Global Change Research Information Office
GCRIO User Services
2250 Pierce Road
University Center, MI 48710
(517) 797-2730
(517) 797-2622
help@gcrio.org
http://www.gcrio.org

Organizations

Society of Protozoologists
c/o Dr. Mark Farmer
University of Georgia
Athens, GA 30602
farmer@emlab.cb.uga.edu
http://www.uga.edu/~protozoa

The Nature Conservancy
1815 North Lynn Street
Arlington, VA 22209
(800) 628-6860
http://www.tnc.org

North American Association for Environmental Education
PO Box 400
Troy, OH 45373
(937) 676-2514
http://www.nceet.snre.umich.edu/../naaee

World Resources Institute
1709 New York Avenue, NW
7th Floor
Washington, DC 20006
(202) 638-6300
http://www.wri.org

Worldwatch Institute
1776 Massachusetts Avenue, NW
Washington, DC 20036
(202) 452-1999
http://www.worldwatch.org

Websites

Agricultural Research Service (USDA)
http://www.ars.usda.gov

Ask a Biologist (Arizona State University)
http://lsvl.l.au.edu/askabiologist/research/index

Amazing Story of Kudzu
http://www.sa.ua.edu/brent/kudzu.htm

Biocontrol Science and Technology
http://gort.ucsd.edu/newjour/b/msg01747.html

Brown Tree Snake Fact Sheet
http://www.pwrc.nbs.gov/BTREESNK.htm

Brown Tree Snake Home Page
http://reorg.nbii.gov/browntreesnake

Canadian Wildflower Society
http://www.acorn-online.com/hedge/cws.htm

Centers for Disease Control and Prevention
http://www.cdc.gov/cdc.htm

Designing Investigations with Pillbugs
http://www.udel.edu/msmith/pillbugs.html

Government Environmental Agencies Links
http://www.studorg.nwu.edu/seed/envirogov

Ecological Society of America
http://www.sdsc.edu/projects/ESA/esa.htm

Environmental Organizations Web Directory
http://www.webdirectory.com

Environmental Systems Research Institute
http://www.esri.com

Exotic Pest Plant Councils
http://www.fleppc.org
http://www.igc.apc.org/ceppc/index
http://www.webriver.com/tn-eppc
http://www.state.va.us/~dcr/invlist.htm

Exotic Plants—Cheatgrass
http://www.npwrc.usgs.gov/resource/othrdata/explant/bromtect.htm

Fish and Wildlife Information Exchange
http://www.fw.vt.edu/fishex/wwwmain.html

Green *Anoles spp.* Home Page
http://www.sonic.net/~melissk/anole.html

Guam USA, Official Website
http://www.gov.gu/

Hawaii State Government Home Page
http://www.state.hi.us

Introduction to Bird Taxonomy—House Sparrow
http://www2.fwi.com/~moellering/taxon.html

Invasive Plants in Florida—Water Hyacinth
http://aquat1.ifas.ufl.edu/hyacomp.html

National Agricultural Pest Information
http://www.ceris.perdue.edu/napis/pests/index

Native Plant Organizations
http://www.wildflower.org/native1.html

Nature Conservancy Exotic Species Abstracts
http://www.tnc.org/science/src/weeds/list.htm

U.S. Congress Office of Technology Assessment
http://www.ots.nap.edu/index.html

U.S. Fish and Wildlife Service
http://www.fws.gov

Weeds Home Page
http://web.css.orst.edu/Topics/Pests/Weeds

World Health Organization
http://www.who.org

Student Evaluation Sheet

Please take a few minutes to complete this student evaluation sheet. It will demonstrate what you thought of the course of study you just completed, and will be used to help fine-tune it for future classes.

1. How well do you think you met the activities' objectives? Explain:

2. Name three of the most important things you think you learned from this course of study.
   a.
   b.
   c.

3. Is there anything you think you could have learned but did not? If so, what?

4. When you or your groups compiled lists of resources to use in your research, what sources yielded especially useful information? What sources did not?

5. How well did your groups' members interact with each other as you proceeded?

6. How would you change this course of study to help you learn more, especially about those things you think you could have learned but did not?

7. Did you enjoy this course of study? yes no (circle one)

NSTA: Committed to Science Education Excellence

The National Science Teachers Association is the world's largest organization dedicated to improving science education on all levels—preschool through college. NSTA promotes this goal through a range of professional activities.

A DEDICATION TO EXCELLENCE Membership in NSTA brings educators into a vibrant organization of more than 53,000 science teachers, science supervisors, administrators, scientists, business and industry representatives, and others representing every facet of science education. NSTA conducts national and regional conventions that attract more than 30,000 attendees annually.

NSTA provides a variety of programs and services for science educators, including awards, professional development workshops, and educational tours. The association is active in promoting the National Science Education Standards. NSTA's "Building a Presence for Science" seeks to align science teaching with the standards in every school in the nation.

SPECIAL PUBLICATIONS NSTA's Special Publications Division publishes books and other media on all key topics and disciplines in science education—including global environmental change, the science of HIV, Earth science, classroom and laboratory assessment, chemistry, physics, and space science.

NSTA members receive notification about the publication of new NSTA books. Members also receive a 10 percent discount on all publications in NSTA's Membership & Publications Catalog, which includes more than 200 specially-chosen titles. The association provides free copies of the catalog and the *Supplement of Science Education Suppliers* to its membership.

JOURNALS & PERIODICALS NSTA is a major publisher of materials for educators. NSTA members choose from among four journals, each for a specific grade level: *Science and Children* (elementary), *Science Scope* (middle level), *The Science Teacher* (high school), and *Journal of College Science Teaching* (college). NSTA journals feature lively, "how to do it" articles, commentary, research, colorful posters, and monthly columns.

Two NSTA magazines designed for students, parents, and teachers—*Dragonfly* and *Quantum*—are available by subscription. *Dragonfly* is designed for use in grades 3–6. *Quantum,* which focuses on math and science, offers feature articles, olympiad-style problems, and brainteasers. NSTA's newspaper, *NSTA Reports!,* features news on education issues, teaching resources, funding opportunities, and more.

Address: 1840 Wilson Blvd., Arlington, VA 22201-3000 USA

Telephone: 1-800-830-3232 (Membership); 1-800-722-NSTA (Publication Sales, North America); 703-243-7100 (Main Number)

Email: membership@nsta.org (Membership); pubsales@nsta.org (Publication Sales); spubs@nsta.org (Special Publications Editorial)

World Wide Web: http://www.nsta.org

MEMBERSHIP BENEFITS

- Award-winning periodicals
- Deep discounts for conventions and other activities
- Advance notice of NSTA conventions, conferences, and publications
- Access to over 30 awards
- Certification programs for teachers in elementary through high school
- Group rate insurance plans
- Voting privileges for individuals

POPULAR NSTA TITLES

- Project Earth Science series, including Astronomy, Geology, Meteorology, & Oceanography
- Views of the Solar System CD-ROM
- Pathways to the National Science Education Standards
- Global Environmental Change series, including Biodiversity, Deforestation, and Carrying Capacity

NSTA JOURNALS & PERIODICALS

- *Science & Children* (elementary)
- *Science Scope* (middle level)
- *The Science Teacher* (high school)
- *Journal of the College Science Teacher* (college)
- *Dragonfly* (students in grades 3-6, by subscription)
- *Quantum: The Magazine of Math and Science* (students, by subscription)
- *NSTA Reports!* (6 times annually)

Global Environmental Change: Biodiversity

Biodiversity uses Costa Rica as a case study in balancing economic growth and resource conservation. The volume introduces students to basic scientific themes and equips them with tools to increase their understanding of biodiversity. Cumulative, field tested, hands-on classroom activities teach students how to integrate science with other disciplines so they can gather information, make decisions, and solve problems. *Biodiversity* is the first installment in the Global Environmental Change series, created by NSTA and the U.S. Environmental Protection Agency. The series covers a broad range of topics pertaining to the impact of human activity on global environmental systems.

Grades 9-12, 1997, 64 pp. **PB# 138X01**

Global Environmental Change: Deforestation

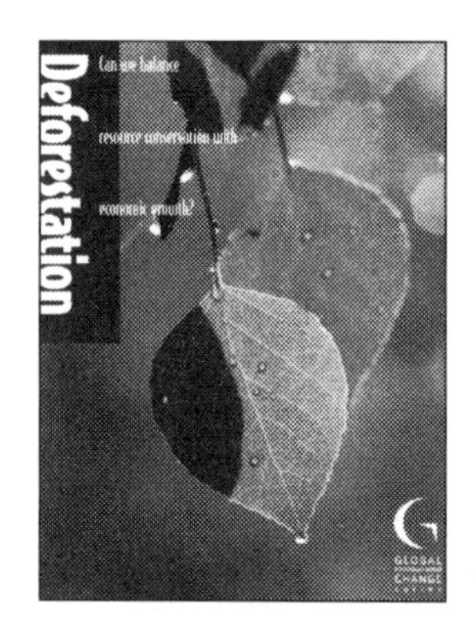

Washington State contains some of America's last remaining old-growth forests. The state's timber supports local, regional, even global economies. Its timber also supports many biological species and provides an important link in biogeo-chemical cycling. How can these roles be balanced? The hands-on classroom activities in this book provide a model for balancing deforestation's ecological and economic impacts. *Deforestation* doesn't hand students answers, but it does provide them with the science-based skills and tools needed to integrate inform-ation derived from a broad variety of sources.

Grades 9-12, 1997, 64 pp. **PB# 138X02**

Global Environmental Change: Carrying Capacity

The science behind the ecological principle of carrying capacity holds true for all species within all ecosystems. These cumulative, hands-on activities effectively demonstrate to students that the carrying capacity principle governs the relationship between species Homo sapiens and system Earth just as it does the relationship between all species and the systems that support them. *Carrying Capacity* uses your class, school, and community as a case study for students to address the question: Can we balance resource conservation with population growth? Carrying Capacity locates scientific activity in "the real world," enabling students to learn by addressing actual environmental challenges.

Grades 9-12, 1997, 64 pp. **PB# 138X03**